THE REAL ATLANTIZ REVISITED

By

Raymond S. Hislop

Shield Crest

ISBN: 978-1-915657-74-9

MMXXIV

A CIP catalogue record for this book
is available from the British Library

Published by
ShieldCrest Publishing,
Boston, Lincs, PE20 3BT England
Tel: +44 (0) 333 8000 890
www.shieldcrest.co.uk

FRONTISPIECE

Photo shows as YYPRAAN GYDRAN KANZAAL wearing ATTYFF, TRIPLE GALAXIES, CORDED, FLYFLOTT cover THNGME, BLACK for closed Galaxy, left facing departure insisted, the sunglasses representing the mask. AT FORMAL for MR DRAGON ZYR and HUMMINGBIRD, ASSESSMENT COMPLETED.

Confirms that ATLANTIZ may be regarded in system at .500 Ag 3d. LOCK DOWN confirmed for 5,101 years on 11:150 review.

Contents

Preface

When I wrote my first Atlantis Text, The Atlantis Hypotheses, it was on the basis that I was dealing with a hidden reality, and different from the Atlantis of Legend. I have come some way from that, but I am reminded that in the prior six books, while they may be viewed as fiction, the intention was to deal with the hidden reality. And there is no easier way to hide the reality, than to do so in plain sight. Fundamentally that is exactly what was and is being done.

I had decided to look again at the Plato text, and acquired a copy. While the commentary on the text was in English, the main text was in Greek which I could not read. I require translations. I had thought that Plato with his circular City with its canals was referencing a Galaxy with its canal connections, but the text and commentary look to be a political discourse, with no understanding of the Galaxy position. While there was a reference to Egypt, it did not appear that the Galactic Interest with Egypt was appreciated and that in Plato's Time the Galactic understanding had been well hidden. If Plato's Atlantis was beyond the Pillars of Hercules, the most likely model was the Canary Islands. The front cover is a photo taken at Tenerife. The other, and more likely, was that Plato had made up what he regarded as an ideal society, albeit one destroyed by the Gods.

I then turned my attention to the Nordic Legend, particularly as they had the word "Yggdrasil". It is similar to the term YYGDRAAZAL which I use meaning this Galaxy, as a concealed term for ATLANTIZ_DRAAZ. The spelling was different, thus highlighting a lack of correct Galactic Interest. This was, after all a

"Flat Earth Society" even if the various "Gods" had a Galactic feel to them.

Given that I am a child of the Reformed Christian tradition, I am aware of the "Burning Bush," and I view that as this Galaxy, and Moses was taken up the mountain to see it, but could not grasp the understanding, or did not want to understand. It is relatively easy for me to see this explanation of the "World Tree" as simply a description of this Galaxy as "The Burning Bush", particularly with the explanation that it burns and burns and does not consume.

I insert here one of my Stylised Star Charts.

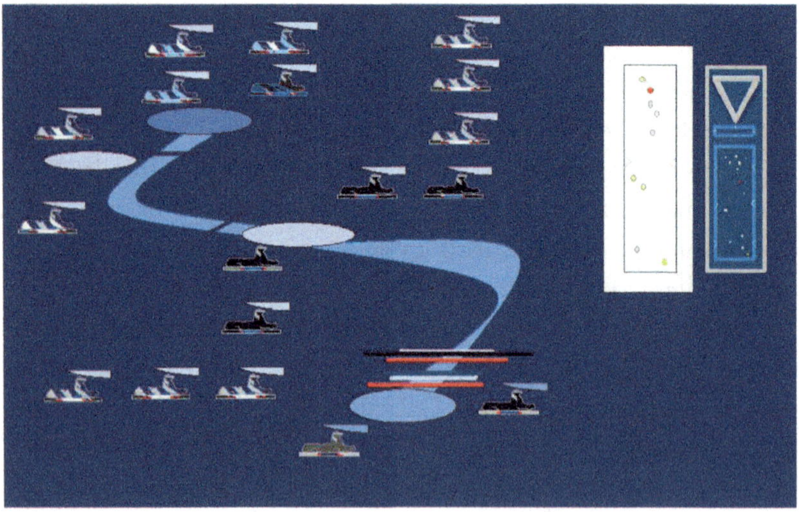

Again to say that the world tree joins the NINE worlds, fits in very neatly with a connection to other FOOTSTOOLZ within this Galaxy, and my Star chart does loosely resemble a tree with branches, with the dark Grey Sphinx at the bottom being the starting point here at ATLANTIZ. The red black and blue lines across the path indicate Gate barring to travel, and there are more "Resting Tyger" than "Nine Worlds"

I am a child of the Reformed Christian Tradition, and it is only that reformed tradition with its insistence on Truth, with currently a more tolerant understanding, which has allowed me to write my Atlantiz texts. That tolerance has evaporated, but the reality of the Galactic ATLANTIZ, with its "Z" marker has been disclosed. My understanding is that of those TWENTY FIVE RTTUR who control the entire Universe, with their respective Galaxies. To a certain extent like the Greek, Roman and Egyptian Gods, where the teaching tool aspect of these "NY TY R", to use the Egyptian, term has been lost.

My block text highlights the Galactic interest.

THAT is not something which in this timescale the local populations appear able to understand either. Thus was it in the beginning, is now and is likely to continue for the foreseeable future as these vast civilised Galaxies and their very advanced Footstools APPLY a LOCK DOWN.

While it is their LOCK DOWN it is applied locally with dogmatic thinking, which has no basis in Galactic Truth.

It was not a stupid question Pontius Pilate asked about Truth, it was just that he was not capable of understanding the reality, and he chose to bow to the mob, who were calling for a blood sacrifice. That was the test to the local population, and it can be clearly seen that the local population here insist on "The Murder Kult"; applying it in the judicial style of Pontius Pilate.

The very advanced civilisations I am dealing with here do superlight travel, with matter transfer. The Kult of the Murderer or The Murder Kult is fundamentally not acceptable to them. At a basic level with the Resurrection of YYGZYVZ, their tremendous power is displayed, but in a very concealed manner. It is all in the New Testament text, if an enlightened approach is taken to that text; again something which appears very difficult for the current local population to do! After all millions are Anti-Khryzt, even in the loosest sense.

There is still currently the demand for a miracle, but after the KAZTDATA meeting in 1952 AD 107 at DHAAL in THRENZ it was specifically agreed that the KURALAKYZ, if visiting, would not be able to use the matter transfer equipment to do any miracles locally. The reason for that was that the ability to do the miracles had contributed to the Judicial murder of YYGZYVZ.

My different spellings highlight the YYG aspect. The child so preciously given, was a visiting Galactic Official, running the "Thief in the Night" and "play with my friends", including the "up close and very personal" Galactic tests! That is a 70 year age Galactic Assessment, and a 35 year or less age is a serious fail.

Pilate's very bad mistake had and has had very serious consequences over thousands of years, and which leaves as the reality this location ATLANTIZ, LOCKED OUT of the Galactic system!

The TWENTY FIVE RTTUR, do not do forgiveness, although they can do selective amnesia, which could be said to be a form of Grace.

When the selective amnesia fails, in the style of Daniel 12:12, they allow a period of 1335 days to expire before delivering a "SLAP DOWN"; a bit like Thor's hammer. Such a breakdown resulted in WW1, and WW2 and the brake to avoid WW3 hangs by a thread, much as it did in August 1945.

It was the local LANAZT Star as MR DRAGON ZYR and HUMMINGBIRD(the former male, the latter female) who asked to be re admitted to THE KHRYZTAAL GALAXIEZ and that has been done simply as I use the name ATLANTIZ and correctly explain what it means in Galactic Terms.

There is one other thing I had to do, and that was to display the DRAGON GOWN and ATTYFF AT FORMAL. Thus the frontispiece photograph and brief commentary!

Even with the weighting at .500 Ag 3d the existence of the gown and ATTYFF is critical to being in the KHRYZTAAL GALAXIEZ system. That is what the local LANAZT star asked THE GREAT COUNCIL to re assess.

That assessment, as is well known to the local LANAZT star turns on exceedingly minor matters, but the fundamental is TRUTH. So I let the DRAGONGOWN with ATTYFF be seen in the photo. It looks exceedingly silly, but is enormously technical.

The GREAT COUNCIL simply asks if the gowns as displayed can be safely worn on the FOOTSTOOL. It is a GYDRAN assessment.

It may take 70 years, it may take 3 or 4 generations of 70 years to complete, but it cannot begin unless the GREAT COUNCIL question can be answered as KONTENT. It is their assessment, not mine. I am well aware that RTTUR MNZYVZ, is pointing out a technical NOT KONTENT, again referencing the Pontius Pilate matter.

In this time frame it is the United Kingdom which is used as the test location; most particularly the territory to the North of Hadrian's Wall. The local LANAZT Star is monitoring, and knows PREZAKTLY the point MNZYVZ is making to MR DRAGON ZYR about the blood sacrifice of THE LORD THE KHRYZT YYGZYVZ.

I describe it as " The Murder Kult." I could call it "Cancel Culture".

The United Kingdom says it allows free speech, but as I have to use the words RESERVE AND REDAKT, the GREAT

COUNCIL is well aware that the marking globally is NOT KONTENT INSISTED and the 5,101 year lock down is APPLIED; 11:150.

MR DRAGON ZYR indicates that the 11 years runs out in 2030, but the point is that EVENTZ mean that that period must be brought to an end early.

This is not a Global assessment, but a Galactic one. The .500 Ag 3d has been very generously applied, and all TWENTY FIVE RTTUR are reluctantly agreed to that. But as their KURALAKYZ is threatened, the KURALAKYZ must now be removed. Departure preparations are in hand. Rather unfortunately for the local population, the KURALAKYZ is insisted withdrawn immediately as certain events, should not have been permitted by the Local Star; thus this brief note in this little book.

Staying with the old Legends and Myths, the "ygg" in the word Yggdrasil, and the Nordic reference to the centre of the "dark world" "ever churning" and sounding like a "roaring kettle" is a pretty fair description of the centre of this ATLANTIZ_ DRAAZ Galaxy, where the central motor spins the Galaxy, and the Nine Worlds could easily be a number of FOOTSTOOLZ near the rim of any of THE KHRYZTAAL GALAXIEZ, which from a distance looks like Galactic dust.

Once again, the Truth and the Galactic Science, neatly hidden in what appears to be a series of fictional tales! The reference to a GIANT or DRAGON, is as far as I am concerned, a Star or Galaxy reference.

From my Star Chart, the upper segments could be said to be the "highest branches" of the World Tree, because the top Galaxy disc with its many Sphinx (Resting Tyger being a Galactic Designation) is an enlightened Group of FOOTSTOOLZ.

The Nine Worlds are given names, and one of them, this world is given the name "Midgard". The trouble with the names is often with the pronunciation, as that can be used to hide the Galactic detail.

I have no knowledge of the Nordic languages, and I if change the name to the Atlantyon Language, "Midgard" would come out as, MY D G AR D. "MY" is "five", " D G", could be used instead of "YYG", and "AR" is the Local LANAZT Star, with the last "D" simply for emphasis. It does look as though the Galaxy Style has been woven in, in a concealed manner and dressed up with the local tales.

Galactic Science deals with reality, not fiction, so I am entitled to ask for Proof of my understanding. I thought I was at the NOT PROVEN, but given that the local population wish to do away with that verdict to secure more convictions, particularly on sexual matters, The GREAT COUNCIL, THE TWENTY FIVE RTTUR, including the one who says "I AM THAT I AM" is entitled to say the matter is PROVEN, and on the basis of miscarriages of justice here to Remove the Galactic Mandate, which the locals did not understand was with them.

As with the Middle Kingdom that Was, with Egypt, and others, the RTTUR can easily do that, and access to Galactic Technology is lost.

The ATLANTIC is the Star Ocean for the LANAZT star. Travelling "from out of Y2GYPT", by way of the Pillars of Hercules, to Mexico,(AZTEKS) (INKA) and (MAYA) and then by way of the North Atlantic Drift to the land of the "Frost Giants" "GRVVNLAND" and then to ICELAND, N0RGE and ZVERGE. That makes EIGHT Realms, but I must not forget the Galaxy Grouping of KATHRYN THE GREAT, which makes NINE.

The travel is by way of Corryvreckan which is in the N0RGE Realm and of THE LORD GUARDIAN, YYG, which is the FVREKKRAN identifier for PYKAAZ PYNWHAAL.

This description is a DRAGON puzzle, and the Realm best known for Dragons is The Middle Kingdom that Was; marked off by a WALL.

The correct list is important, for it is like a travel plan to disclose the correct understanding and route for Galactic Travel and communication.

Glasgow to London to Perth (Western Australia) and then to Auckland is a Galactic marker route, because with aircraft(safe civilian air travel is a Galactic Marker) it discloses various Societies who have at least a basic understanding of the Galactic requirement; that "nothing which is hurtful or harmful will be admitted." I must check the Isaiah Text for that comment.

Newton, as I have previously explained, could not find the Bible Code, because in the time he was living the Galactic PROTOKOLZ were not met.

Are they met today? METHINKS NOT INSISTED!

I RESERVE AND REDAKT.

The Milky Way means "male" and "female." FINAL ANSWER! The NYKYDYNYZZ material,(wrongly called dna) is the design specification and correct design is material for use of the matter transfer equipment and for much of the exceptionally advanced technology. Thus the GALACTIC PROTKOLZ are fundamental.

In this book I am using silver (Ag) coins of certain purity to describe the various Globes within any of the PP

(PYKAAZ_PYNWHAAL) Galaxies, with as of 2024 AD 107, this Globe being 3d Ag 500.

The minimum weighting for a globe to be IN SYSTEM is 3d Ag 925, so this one falls short.

For various reasons the Galactic weighting is at 3d Ag 500, which is below the level for establishing Galactic Connections.

I am aware of various Science Fiction series, but because of restrictive copyright legislation, I am restricted in what I might say of such. What I can say is that these very advanced civilisations can look at such and can set Galactic Science questions on the material shown. The fancy DRAGON GOWNZ and especially the fancy ATTYFF headdress are not understood. The nearest, is the ancient Egyptian Atef crown, and the various others as the locals struggled to understand the complex meaning. In Mexico, likewise in other parts of the Americas, there was only a vague grasp of the ATTYFF which is a headdress of specific understanding, light, not heavy, and with the FLYFLOTT at the rear, the FLYFLOTT shape and colour depicting the Galaxy to which the wearer has access, and with TWO feathers which may be worn, not more or less. When the feathers are not worn, they may be said to be invisibly worn, as might designate a VISIBLE INVISIBLE.

Harder to understand were those Galactic Individuals who wore only the cover, marking a closed Galaxy, and with the double points representing specific and powerful Galaxies, it very much looked like a serpents tongue, which it most certainly is not. To add to the confusion the visitors referred to one of their advanced travel mechanisms as "The Double Headed Serpent" and the pyramid in the Americas has at certain times the appearance of two serpents on either side, with an individual emerging from each mouth.

Hidden in plain sight, the Galactic understanding and their visitors!

Once the Spaniards conquered the Americas, and imposed vicious Roman Rule, all the Galactic science held by the locals was lost.

So I have provided this update, as the REAL ATLANTIZ is REVISITED.

Superlight capable civilisations weigh in at 1/- Ag 925+, with correct matter transfer and transiting they weigh in at 2/6 Ag 925+ and at the super advanced level, and who have been in that state for thousands and millions of years 5/- + Ag 925+.

These are Galactic assessments, NOT local, and the FLYFLOTTZ for a system Galaxy weigh in at 10/6d, which can be 3 pieces of Ag or more, including 30! [28 x3d =7/- +1/- +2/6d]

I insert here a photograph of some plates. These represent the Galactic understanding at an advanced level, the plates representing the Galaxies as seen, the Coins the FOOTSTOOLZ.

This white bowl represents a Galaxy as seen through a telescope. A large, mainly white Galaxy! The 3d coins represent their footstools. They are not sufficiently advanced for superlight travel and communication. Collectively they weigh in at 16 x 3d =4/-, which is below the 10/6d weight for a compatible Galaxy with superlight travel and transiting.

This photo now shows coins to a face value of 10 /-, but one of the coins is only Ag 500, so that this particular Galaxy of the PYKAAZ PYNWHAAL style, as the design of the bowl and mats designate, is only 2 x 2/6d, 1 by a shilling and 1 x 6d (2 x 3d together in one place) and 4 x 3d. It does mean it has two which have superlight and transiting, one which has superlight travel but no transiting, and at a total value of 7/6d, the Galaxy does not reach the 10/6d for a System Galaxy. I am pressed to increase the weighting here at ATLANTIZ to 3d Ag 925. I am not willing to do so, because in the Territory of The Lord Guardian, someone has deliberately stamped on DRAGON toes.

DRAGON is in KREAM PUFF, and is inclined to Spit Fire at those who have so insulted him. Dragon is only slightly and carefully restrained to keep in the Divine wind, because after all in terms of Daniel 12:12 the 1335 days have expired. MR DRAGON ZYR knows they have expired, but is persuaded to a RESERVE AND REDAKT. He is counting Activators. They are the consents from the 25 RTTUR. Only one is holding out, and as to be expected it is THNGME.

RESERVE and REDAKT.

I am going to use the Galactic Style, as previously, Preface, Old Text, New Text and Technical Specification.

27 March 2024 AD 107

Raymond S Hislop

RTTUR AM AMEN

The Old Understanding

≈ 1 ≈

Smoke Signals to a Supercomputer

I had used that expression in one of the previous books pointing out the considerable gap in understanding between different societies.

Currently those searching for Aliens are looking for intelligent life, symbols and messages such as the "signal relating to hydrogen". The prospect of ultra advanced Aliens is not really covered, except to say that there is no sign of such. I could add here that they are looking for the wrong aliens, and with the wrong equipment.

In applying the technical name ATLANTIZ I am making the point that I am describing very real Aliens who are able to see clearly what is happening on this ONE of THEIR FOOTSTOOLZ, but one which is restricted and most certainly Not a PEARL HARB0R; their description of one of their Globes which has communication and travel links. Thus, as I have previously said, a Galactic problem when on 7 December 1941 Nyppon attacked Pearl Harbor. That did not awaken a sleeping DRAGON, because in Galactic Science terms, the DRAGON who is the local LANAZT star "neither slumbers nor sleeps." He may be Resting but do not stand on his tail, nor poke him in the eye as he rests. Lady dragons? Yes, because the local star has, in Galactic Science terms, two forms, MR DRAGON ZYR, male, and HUMMINGBIRD which is female. On the FOOTSTOOL, male dragon times can be times of war, whereas female Dragon times are times of peace.

After Pearl Harbor attack, the Dragon response came after just over 1335 days, as per Daniel 12:12; and the Divine wind hit two cities. The DRAGON is not the United States, they were merely the vehicle; the Dragon was, is and always will be, the local LANAZT Star.

So it is the local LANAZT star which asks the GREAT COUNCIL of THE KHRYZTAAL GALAXIEZ to be re connected, knowing full well that a LOCK DOWN and THROW OUT has been imposed. With such LOCK DOWN in place, there is no prospect of communication, far less any Alien visitors. The scientists may look though a glass darkly with their telescopes, but all they see at the Galaxy plates, the spinning spiral Galaxies, as with M31, M51 or the few Sculpted ones; what they do not find are the hugely secured FOOTSTOOLZ, because the FUNDAMENTAL of this large Advanced Galaxy Group, is that " Nothing which is hurtful or harmful will be admitted." I find the DRAGONSPEAK words in the Text of Isaiah, so I can be satisfied that the Galactic record has been left and UPDATE only when the King James Bible is produced, and in its English format. I see clearly the Galactic style of Preface, Old Text, New Text and Technical Specification(Revelation with its talk of Dragons, Pearls(Harbours) and of course the request for the return of the Galaxy official YYG. It is hugely unlikely that MNZYVZ RTTUR will permit the return of his YYG ZYVZ, after the mess made by Pontius Pilate. He is persuaded, with the AD 79 eruption of Vesuvius, to leave Rome alone for the present, but all the TWENTY FIVE RTTUR are well aware Rome is unsafe as it is built on the caldera of a volcano. That was the sacred flame the vestal virgins were "protecting".

The volcanoes are activated by the local LANAZT star. That is the Galactic Science point, as also the physical Geography point. They do not simply go off. There is a reason why they go off, most usually because THE DRAGON (a short Reference for MR DRAGON ZYR, the local LANAZT STAR) has been annoyed, and some rectification is required. All the Atlantic Volcanoes, which includes such as Vesuvius, and those near the Black Sea are indicative that a particular DRAGON, (and by that here I mean a Galaxy grouping) is angry, very angry and wishes MR DRAGON ZYR to deal with the matter by way of rectification. Such rectification can be very messy, as with WW1, WW2 and WW3. The latter is in the hands of MR DRAGON ZYR, because someone locally has done something they should not have done.

I have used the numbering R18, adding sometimes, "reserved in family, insisted but not forced." The R18 label on films means they are restricted to Adults for viewing and with a sexual content, but generally it makes the age of 18 the dividing line between Adults and Children. Thus certain items on a PROTOKOL basis should not be made available to under 18's. I could thus say guns should not be made available to under 18's. I hear of an incident outwith RESERVED TERRITORY, concerning guns, and used by an individual aged under 18, which is simply not acceptable. I see the volcano in Iceland active, and that volcano looks very much like a Dragon's tail. When a dragon's tail is slapped down there is trouble.

RESERVE and REDAKT

The Pearl Harbor event led directly to a GALACTIC SLAP DOWN after the 1335 days but has not been appreciated as such locally.

Now the Galactic weighting has been APPLIED at 3d Ag 500 so now a new PEARL HARB0UR EVENT is cleared for ACTIVATION. THE DRAGON is not amused. That is ultra polite speak for meaning The Dragon is furiously angry.

The Dragon can deal with such very easily! Repeat, Mount St Helenz or Tambora, even Tamfamaguran on the Atlantic Ridge and shut down the Atlantic conveyor! That would sort out Global warming by far more than 2 degrees celcius!

Now considerable care has to be taken not to overstep THE DRAGON, for at this stage, 2024 AD 107 there is a game of Galactic Chess at play! Again, as with all DRAGON matters it is not obvious, and if the DRAGON is playing, sometimes he plays white pieces sometimes she plays black pieces. Red and White, Blue and White, but it becomes tricky if all white pieces are played or all black pieces; in which case weighting has to be APPLIED.

The Icelandic Volcano sends the smoke signal to the local STAR, which, while it might be said to be a high temperature super computer, is well able to understand that smoke signal, and the request to THE TWENTY FIVE.

In terms of the Hymn Jerusalem, it is a request to "Bring me my Chariots of Fire." Except that the request is not for a Chariot of fire but for a Boat of Millions of years to return the visiting YYG, as the report has been made to the TWENTY FIVE RTTUR.

What John had asked for in Revelation,[22:20] and what the Local Star had asked for had been done, and the Galactic weighting had been applied to the FOOTSTOOL at 3d Ag 500.

The request for a review on appeal was made and it was ARRAN KO ATLZ who marked NOT KONTENT INSISTED. It remained at 3d Ag 500.

The fundamental was "milky way practices, male and female", and to make 3d Ag 925 there had to be safe civilian Air travel in comfort and at reasonable cost, and clearly that was at best "a work in progress."

The TWENTY FIVE RTTUR had arranged to have put in place the Internet, and from the information on that, it was clear that they had been over generous with the 3d Ag 500. The request for aerial photos of the Cities of Kyen and Taydaranakushton, or any other of the 2/6d + ZYON was refused at present.

PROTOKOLZ

As I have done in previous books I am using these PROTOKOLZ as a point of Clarification.

The point being made here is that with an Internet in place, these vast Galactic Civilisations have access to all the information on line. They are thus able to see very clearly that the local populations continue to remain unsafe, and as various incidents make clear many of local population are of The Murder Kult.[TMK], and their territories unsafe. I do a reserve and redact, because these texts are written from the viewpoint of these vast Galaxies, examining this location as to whether they might drop shields and communicate. With effect from Midnight on 31 March 2024 AD 107 the LOCK DOWN, with POWER CLOSE is APPLIED INSISTED.

Various countries can do as they please, but where the Galactic PROTOKOLZ are not met, there is no communication, and such countries clearly are hurtful and harmful, and their individuals cannot be admitted.

While it may thus seem impossible, the RTTUR adopt various positions. Thus the Galaxy M64, LLYKOTT_LLYDRAN may take a more generous view, but then it can be difficult, which is why it is sometimes simply called the <<<LL>>> system; and I doubt if many here would wish to go to <<<LL>>>. Some parts are very pleasant and hugely civilised, others very much not so.

The THUURRNN Group is likewise exceptionally difficult, and as they are a massive group of Galaxies THAT is one Group which must be carefully considered with any admissions and they can easily refer those wishing to enter THUURRNN territory, to the<<< LL >>> system.

I use the triple chevrons simply to make the Galactic point of Triple Galaxies, the group FVREKKRAN, FVHATTHARAN and THUURRNN; THE MIGHTY THREE GALAXIEZ. This is not a maths text, so the chevrons are not read as "less than" or "more than" rather to make the Galaxy point that it is abbreviate for the Triple Galaxies in System, LLYKOTT_LLYDRAN. The underscore, if used makes the point of at least TWO GALAXIEZ and MUCH LARGER GROUP. Whether as with PP LLYAAZ_ATTHENYA they use the PP before the name to make the correct reference to the sign of the "The Fish" and "The Whale", which is to mean a water based Footstool, with large fish and whale designs, to be part of this THE KHRYZTAAL

GALAXIEZ group, the FOOTSTOOL must have whale and fish populations; some of which may be very dangerous. That is a FOOTSTOOL protection identifier.

≈ 2 ≈

Galactic Dragonz

The trouble with the DRAGON term, as with the City term, is that often they mean Galaxies, particularly Barred Spirals. Although not appreciated at this Galaxy, ATLANTIZ_DRAAZ as viewed from other Galaxies, appears as a Barred Spiral, which means to those advanced civilisations which have superlight travel, this Galaxy has what may be compatible species on certain of its FOOTSTOOLZ; as with DRAAZ.

While it may be written in English as ZY0N, in DRAGONSPEAK, that means an Enlightened and peaceful City and one which is in the Group <<<PP >>>. The numeral 0 represents the Galactic Disc, and the letter Z to make the point that it is Galactic. The Golden Fleece is a reference for this Galaxy, as is the term YYGDRAAZAL. The latter is a hugely secured reference for ATLANTIZ_DRAAZ, and discloses an advanced understanding.

BHVAAL_ADDVVR is thus a Galaxy term, as may have been misheard, and may be misunderstood and is reference to the AAGHAADDWR Galaxy (Silverado.) Various spellings are used. The MAYA called it ADWR and that reference can be found on their Glyphs.

PROTOKOL

All THE KHRYZTAAL GALAXIEZ SYSTEM can adjust names; including to shorten or lengthen, as with very formal names. There is a PASSCODE aspect with the names as written and pronounced.

≈ 3 ≈

RTTUR at The Round Table

Within the various KAZTDATA HALLS in the various Galaxies, the table which seats the TWENTY FIVE is oval. The round table is reserved to meetings of THE FIVE [NY MY GURY0N] who consider the appeals on ADMISSION, RE_ADMISSION or THROW OUT, and always on that appeal panel sits RTTUR THNGME. Occasionally, within which ever house he is living, there will be a circular table which could be extended to oval, and which can accommodate the FIVE RTTUR.

The LANAZT star had asked for an appeal from the 3d Ag 500, and THNGME had made it plain he would not reconsider. Indeed RTTUR MNZYVZ was still smarting, and like RTTUR THNGME, they do not do forgiveness, and such should not be asked, BY ANYONE!

What THNGME and MNZYVZ and some of the others were willing to do, was what they referred to as "Selective Amnesia." Such was hard fought, and was sometimes described as Grace. The PROTOKOL was ultra strict, "nothing which was hurtful or harmful would be admitted to their Galaxy Groups". The Seventh or Tenth Heaven was well short of the Twenty Five Galaxy Groups, especially large ones like THNGME which counted over 1,000 Galaxies in their group, all working to PYKAAZ_PYNWHAAL and MILKY_WAY practices.

In making the application in 1952 AD 107, the LANAZT star was well aware that both THNGME and MNZYVZ RTTUR would require to approve, and with that there were considerable risks; some of the risks were known, others were unseen.

The 3d Ag 500 did mean that even with Reservations and Restrictions ATLANTIZ had been re admitted, but there were problems.

The date references BCE and CE do not conform to the ATLANTIZ designation set in place by the YYG, as AD. While it might have been said to be read as "in the year of our Lord", THAT date reference also representing ATLANTIZ_DRAAZ was the RTTUR insistence; all TWENTY FIVE of them. When that was not done the Galactic Gates were locked barred and bolted. There was no way the "double headed Serpent mechanism" could be used. THAT required safe civilian air travel, and while THNGME pointed to one incident in TECHNICALLY RESERVED TERRITORY, RTTUR FVRGHAAL pointed to another. As with the "Titanic" incident both RTTUR pointed to Government Cover ups, and the need for the Local Star to arrange for the matter be brought to light, and the Governments Responsible being made to perform the "Kow Tow." The Lady RTTUR PP HAMA GIGHA was particularly pleased that the submission was to be made "Kow Tow", not to the successor to the Middle Kingdom that was, but rather to the YYG; the visiting Galactic official. THAT in certain respects was relatively easy, but also exceedingly difficult. There would be severe loss of face, but the RTTUR insistence was that if there was to be progress to the 3d Ag 925 weighting THAT was INSISTED. NO KOVER UPS!

The English text belied the Dragon insistence and urgency. There was also the little matter of defectively certified engine parts; particularly the turbine blades, and the lack of understanding of the importance of safe civilian air travel and the safe manufacture of such craft.

In all it really depended on whether the local Star wished to push the button for WW3, or whether the intention was to move up a GALACTIC level.

The list of LOCK DOWN matters was extensive.

Whether the temperature at the LANAZT FOOTSTOOL, increased or decreased was a matter for the local Star; not for anyone on the FOOTSTOOL, with the possible exception of the YYG THNGME

The old local philosophies had struggled with the Galactic Concepts. The MAYA had understood that a HEAVEN was a Galaxy, and if you were travelling to Seventh Heaven, it was to the Galaxy which was Seventh on the list; a bit like travelling from Glasgow to Auckland by plane on a route, London, Perth WA, Brisbane, and then to Auckland. That was only four stops, but the RTTUR INSISTENCE was likely to be DIRECT, and in a rather large flying boat; the so called Boat of Millions of Years. When the locals did not have safe civilian air travel, they were not entitled to see the Real Boat of Millions of Years, with its label on the side, and in MNZYVZ and THUURRNN colours, light blue and white. They were not Unidentified Flying Objects; they were very much clearly identified and marked as Boats of Millions of Years. Their use was to confirm GALACTIC CONNECTIONS and to bring in advanced technology and medicines.

THNGME RTTUR was not at the round table when on that 21 December date he heard there had been an incident where wreckage of a large civilian airliner had landed in THUURRNN designated RESERVED TERRITORY, to the North of Hadrian's Wall. That wall had been built as a marker that nothing which was hurtful or harmful, including Roman troops, were to be allowed beyond that Wall. THAT assurance had now been broken, so potentially that was a mushroom cloud matter. The 1707 Treaty of Union conveniently concealed its DRAGONSPEAK, ZALT aspect!

Transgress the THUURRNN PROTOKOLZ and for such THNGME could easily instruct snap of the DRAGONSTICK, but that was not required. The tidy up was insisted 21 December 2023 AD 107 and the TWENTY FIVE RTTUR marked KONKUR INSISTED. IMMEDIATE.

Different spelling, to make the Galactic point!

The instruction was given to FVRGHAAL to find the aircraft wreckage and expose the deceit. If necessary that would include matters RESERVE AND REDAKT; a matter which the TWENTY FIVE trusted could be avoided, with the performance of The Kow Tow to The visiting Galaxy official.

The use of nuclear weapons was a matter for THE TWENTY FIVE, and MR DRAGON ZYR or HUMMINGBIRD and not for **any earthly power. Such powers only held the Dragon's Teeth to** await the instruction from MR DRAGON ZYR, despite what any locals might think!

PROTOKOLZ

The difficulty with what is written here is that it has to incorporate the DRAGONSPEAK and Limited Disclosure on a RESERVE and REDAKT basis.

Despite what the local politicians in the UK might say there is no free speech here and MNZYVZ has pointed out to THNGME that the specific matter with the legal murder of the YYG ZYVZ has not been resolved; **far from it in fact!**

The DRAGONSPEAK is to draw attention to the fact that the **Nuclear weapons are referred to as Dragon's Teeth,** as also to make the point that the expression **"the Divine Wind"** is for the Mushroom Clouds the use of which and the authorisation has to come from the Local LANAZT Star.

The use of such is not wanted, but clearly the two aircraft incidents have been selected as ACTIVATORS; but there have been many more.

The real point here is that the TWENTY FIVE RTTUR wish some disclosure of their interest, and that is because of the 3d Ag 500 weighting now applied.

It is with Events that the DRAGON handiwork is seen. The invasion of the United Kingdom by individuals on the rubber boats and the desire of some the local politicians to divide the United

Kingdom has, in Galactic terms, gone down like a ton of plutonium bricks. These Galaxies are well able to do a SLAP DOWN, as with the Spanish Flu, and take out thousands of millions globally who do not conform to their design requirements. This is not easily appreciated. Neither is it appreciated that KOVER UPS are likely to mean a GALACTIC SLAP DOWN will be INSISTED.

My block usage is to make it KHRYZTAAL KLAAR that there is a real Galactic position, and persistent transgressions will lead to SLAP DOWN, as with WW3!

To travel in this Galaxy Group, individuals must meet the PROTOKOLZ of the Galaxies to be visited, and some of these advanced GALAXIEZ will simply refuse those who do not meet their designs; with fatal consequences. This is critical, as even one of the local Science Fiction series disclosed at the outset. It is unlikely the local producers appreciated the technical aspect, but certainly the Galactic eyes would.

This applies particularly to the styles of Aliens. FVREKKRAN, the Whirlpool Galaxy, has ultra strict criteria for the approved populations head shapes, which must be to the four "beast" styles, Atlantyon, Wolf/Dog, Tyger and Rhino in the bi pedal design. Head shapes are super critical. The Xolo dog is a living example of Anubyz style, but to conform to design specification, the head has to be on the bi pedal tailless Aliens. Aliens even as actors, who do not conform to the FVREKKRAN design specification, do not meet the technical specification and put at risk even the 3d Ag 500 weighting; particularly as with current film technology, such Alien designs can be done. This is one of the areas where the ATTYFF headdress and DRAGON gown are hyper critical. I have disclosed the designs, so in essence they can be worn by individuals Acting, but as said previously even with actors, it is only safe for those baptised, trinity style, to wear such. MR DRAGON ZYR or HUMMINGBIRD will know if those wearing even as Actors are entitled to do so, and the Baptisms, are a MR DRAGON ZYR matter. A point which does not appear to be correctly appreciated locally!

There is thus now a LIMITED LICENCE for Actors to wear the DRAGON GOWNS with ATTYFF, or for Actors to have masks made to resemble the TYGER, DOG/WOLF and RHINO style. Do remember that FVREKKRAN are hugely critical of correct head designs; even with KARTOON characters or with computer designed individuals. Real bi pedal aliens DO NOT have strange ears or antennae, and certain styles as with insects and reptiles are not acceptable. This is HUGELY technical in GALACTIC SCIENCE TERMS. The Rhino style incorporate such as the Z0BYK designs so rough skins are acceptable. The Egyptian Sobek is shown WRONG DESIGN. Head shape of KAIMEN on quadruped body is fine, but substantial redesign is required for bi pedal.

The point here is that if aliens are produced for films, they can be anything the producers wish, but make aliens in the Tyger, Wolf/Dog. Rhino style of the actual real aliens, it must be appreciated that the real Aliens would be indistinguishable, and so could be placed into such films. Again a very important GALACTIC SCIENCE POINT!

RTTUR LLOMOTTOR has spotted Rhino styles, which are very similar to his species, and is not pleased at the misuse. Z0BYK style is also of interest at FVRGHAAL [The Great Southern Pynwhaal Galaxy.]

To show fake aliens from alleged crashed UFO's is to demonstrate a complete lack of understanding of the REAL Galactic Science.

The nearest System Galaxy is M 31 THRENZ, so for them to travel here, their technology is extraordinarily advanced. KYEN is a large city of the Southern Continent, like Australia New Zealand combined, and KYEN individuals have visited here, and while locally here they would generally take ATLANTYON style, not their WHITE TYGER style.

Thus in Atlantyon shape, if asked where they had come from, might say "from the west", which would be correct, but would be misleading in that they would have travelled millions of light years, not from a landmass nearby.

The ATTYFF headdress and DRAGON GOWN are INSISTED in this Group, as with the face masks(including sun glasses) in a Formal Setting. The facemasks, including the fancy plastic KHAAR NY FVHAAL masks make the point of the VISIBLE INVISIBLE and HOSTING aspect. It does not mean that an individual wearing such is actually hosting, but as an ACTOR can make the hugely technical and many Galaxies point and the VISIBLE INVISIBLE understanding. The masked photographs I have produced are making an ACTING and HOSTING point.

The wearing of Carnival masks does not make a territory RESERVED; because Venice as part of Italy is a NOT RESERVED location.

Artificial intelligence, which appears as current local flavour, this Group of Galaxies only do to a limited extent, and their insistence is on MILKY WAY, R18 Reserved in Family, insisted but not forced. In a 3d Ag 500 globe **changing gender cannot and should not be done**. THAT is FINAL! It contravenes the MEDICAL PROTOKOL and the HIPPOCRATIC OATH, and WILL LEAD to serious medical problems. It discloses to these Advanced Galaxies that the Doctors are NOT correctly taught and are practising the wrong medicine. Effective antibiotics are likely to be withdrawn!

RTTUR THNGME makes the point as does RTTUR FVRGHAAL.

It is very easy for THNGME to remove the Mandate of Heaven from the United Kingdom, especially when the United Kingdom failed to understand and appreciate that they held it; a matter with which the LANAZT star is entirely familiar.

YYPRAAN GYDRAN KANZAAL

≈ 4 ≈

Selective Amnesia

RTTUR MNZYVZ was fully aware that RTTUR THNGME would have none of it, and his selective amnesia was rapidly failing. Memory was returning, as he decoded the concealed information in the Matthew text.

MNZYVZ was aware that ATLANTIZ was primarily a THUURRNN footstool, and while the locals might call it Earth, such spelling hid the THUURRNN , YYRTH.

The locals could talk in terms of a "virgin birth" but such designation simply masked the Galactic Interest. The 2/6d and above weighted footstools could easily do the matter transfer of the milk to allow the female to become pregnant, but in THUURRNN terms they would insist on the "old fashioned way"; male and female, and they would not approve of Artificial insemination. Yes technically it could be done, but that was not the point.

The star in the night sky represented one of their visiting Boats of Millions of years, and in that case one of the "three ships" had been THNGME. The position had been checked with THUURRNN who were still most displeased and had withdrawn. They had no objection to him selecting an individual to run the monitoring exercise, like a thief in the night, and as such the YYG was put in place. Taking part MNZYVZ name, the male child was to be called YYGZYVZ, with pronunciation "Gesus". THUURRNN would naturally continue to monitor, and given local sensitivities, THUURRNN suggested that shortly after birth he should be taken into Egypt, where both THRENZ and THUURRNN could consider imparting the necessary information so that the 70 year test might be run. THUURRNN was doubtful, as were THRENZ that it would achieve the result, but RTTUR MNZYVZ was entitled to try.

The three wise men were thus the Galaxiez, THRENZ, THNGME and MNZYVZ, and the messengers announcing the arrival simply the messengers on the Boat of Millions of Years. Very easy for them to appear as visitors from the East delivering gifts of Gold, Frankincense and Myrrh, and with their computer technology allow angelic music to be heard in a local area. It would be very technical if they allowed to be heard TZAD0K the Priest, which Handel was yet to write. The difficulty is that at a later date they could impart to Handel the tune so that he might write. As I have explained previously these system Galaxies do HOSTING, and in that way they can import invisibly advanced technology which is then simply viewed as a discovery. They can also host the advanced technology and then withdraw it where the locals misbehave; their assessment!

The escape into Egypt covers the technical point, that while the appearance is given of circumcision, that was not done to give the child the Egyptian appearance, as also because neither THUURRNN nor THRENZ would impart their technical information to a circumcised individual.

The technical aspect here is with the R 18 Reserved in family. The child is minor until 18 and there must be no sexual interference with such, male or female. This as MNZYVZ knew was not a detail.

So the watching was done over a number of years, and all three RTTUR could monitor the local situation. Other RTTUR could do likewise.

Roman rule was wolf matter which brought in Sombrero[TY MA HY HAAT] protection, and it was made plain to the Romans that care had to be taken with the visiting Galactic individual. Their visiting official was not a prophet, but was a high ranking Galactic official, running Galactic checks.

The opposition to the visiting Galactic official was noted, as his Galactic attitude and understanding conflicted with the local position.

MNZYVZ was well aware of the fall out with Egypt, from whom they had withdrawn the so called "Mandate of Heaven." The TUAT ANKH AMEN.

Different languages have different spellings, but in writing this I am required to reflect the correct Galactic understanding. So it was entirely correct so say that YYG ZYVZ was not a prophet, because now with the Galactic input his home was light years away; his work locally was still being done; the purpose of the exercise was to explain the serious errors in Old Testament understanding and in particular to reject the Murder Kult. However as time passed it was apparent the locals would not give up The Murder Kult, and despite TY MA HY HAAT, trying to deflect from it, at a crucial point Pontius Pilate failed, to the furious anger of TY MA HY HAAT and the other RTTUR. For that TY MA HY HAAT would punish Rome, although that would take thousands of years. First they would build it up, then they would slap it down; and smash it to pieces. Generally when buildings and Temples are smashed it is to make the point that the unseen Galaxy connection is broken.

With the judicial murder of their Galactic official, the matter of punishment was considered. TY MA HY HAAT would deal with the destruction, although that would take a few years. They would arrange to have a particular boundary wall built and marked off to one of their officials with Mushroom clouds displayed. Hadrian was provided with the explanation of their requirement; thus he built a Wall to designate a Galactic Boundary not to be crossed.

THUURRNN arranged to have a DRAGONSPEAK record left in their style, after 1000+ years, so that they could take a view. They confirmed the Wall marker, so that the territory to the North of the Wall was to be designated MNZYVZ, THUURRNN THRENZ, and if the murder kult was brought in there, they would be very vengeful, including authorising the Mushroom clouds.

Warning would be given on their insistence of a PEARL HARB0R designation being required, and transgressions would result in the deployment of two Mushroom clouds; locally delivered. Such would mean that the locals would be allowed to develop such, and authorisation given to the LANAZT Star, only to be used in the designated MR DRAGON ZYR periods.

The book was sealed up, the MY DYRAN, the Galactic Direction given.

This chapter can be checked with the Matthew text for accuracy. I am satisfied that it represents the TTM position 29 December 2023. 70+ has been achieved, and extension agreed. The Activators are in place for two enormous Mushroom Clouds, and/or Two massive volcanic eruptions Krakatao + Eye of a Jackal. That makes Iceland M101, TY MA HYHAAT and THUURRNN territory.

≈ 5 ≈

The Difficult Reality

The difficult reality is, as I said in an earlier text that, "The Truth is obvious. It is so obvious that it is seen only with difficulty, but is seen all the time."

The big yellow disc can be regarded in that light. Applying the name the LANAZT star, means that I am viewing it in an entirely new light. Not entirely new because I have referred to it as MR DRAGON ZYR and as HUMMINGBIRD, as two different descriptions for it.

Before Genesis was written, Amenophis IV had appreciated that, so that in essence Genesis did not convey that Galactic understanding, and instead Genesis sought to project a brutal fiction. Moses was the second son, and by adoption and was jealous of the first born, Rameses, and it is that jealousy which is displayed throughout the Old Testament Text by Moses, with a hatred of all things Egyptian. Later Old Testament Texts as with MY KAAH, AM0Z, HAFVA KVV00K, contain within their headings the correct Galactic detail; and that includes the understanding that "to communicate, brief words are best."

Yes there can be a great flood as the modern geologists should know, because such massive crust movements can change the lush vegetation to arid deserts and can shift mountains, place the vegetation underground, and crush it to form oil. Mountains can tell they were once underground, and water erosion proves that certain arid deserts were large lakes. The geology is factual, but ignored and not appreciated by many of the Global population. That said THE TRUTH does not change, which is why it is THE TRUTH. There are of course the volcanoes, but the Galactic mechanism which activates the volcanoes is not even vaguely grasped. I have referred to such volcanoes as HYHASTERN and others as KULYN; and the styles are that the Hyhastern safely absorb the hot Plasma from the local star, and the Kulyn send it back to the local star, sometimes with minor eruptions, and at other times violent eruptions.

As for the Star plasma, I have referred to that like a buzzing bee, so that locally when matters are relatively peaceful the local LANAZT star may be said to be in "HUMMINGBIRD" mode, and in times of difficulty and War, in a MR DRAGON ZYR phase. Currently in 2023 it has just started a Mr DRAGON ZYR phase, as war erupts in Ukraine and in Israel.

These are Star matters and the LANAZT star, is not simply a big ball of gas, but is a living entity; as are all the other system stars. In clumps they form Galaxies, Spiral, Sy Fhaart and Sculptor. Their very advanced populations are shielded from view, as I have explained with ceramic plates and coins explanation.

Each of the plates shown below represents a different style of Galaxy as seen with telescopes. The square plate is of a Sculpted Galaxy.

The plates with no coins are Galaxies which do not have the correct weighting to be a KHRYZTAAL GALAXY. [They have not reached the 10/6d value.]

The mat on which the plates are placed represent PYKAAZ_PYNWHAAL as does the sculpted system with the light coloured "Fish" and the dark coloured "Whales."

The white bowl with the yellow containers is of a hidden galaxy, as the container can hide the 10/6d plus weighting.

The large plate with the £2.00 coin makes the point that at least one of its civilisations is well above 10/6d, and its Ag weighting is above .925.

These examples are teaching tools, and where such understanding is not available, the weighting locally cannot exceed 3d Ag 500; although it does advance towards Ag 925.

MILKY WAY is a species style, male and female and is a fundamental within what I call the KHRYZTAAL GALAXIEZ. This can be dressed up as "made male and female" because that design is vitally important on compatibility for travel. The Galaxies are designed, as should be obvious with their spiral design, and evolution is an adjustment mechanism, as may be seen with the design of cars, trains and planes.

Those who worship evolution as with Darwin; have the completely wrong understanding. In Darwin's case as with Moses, it is bitterness and anger which takes them in the wrong direction.

Global warming and cooling, as Amenophis IV understood, are matters for the local star. The local ice caps, if examined, as indeed they are being examined, give details of the climate change; the periods of change; further back in time the different ages of the Footstool. All understood in technical local science as with geology. Oil reserves were once forests. Much of the real science does not fit with local Dogma, and there is a serious danger when dogma collides with scientific reality; for it is dogma which has to give way. Thus Flat Earth dogma has had to give way to Global. But it is much harder to move from Global to Galactic as Amenophis IV discovered; really easy to make mistakes when the insistence is only with one brutal dogma. The Gods of ancient Egypt, as with those in the Greek and Roman Pantheon, was a short hand aide memoir for the various different compatible Galaxies and their species. Yes adjustments had to be made to put the correct understanding in place, but to replace it with brutal and murderous dogma simply will not do. Take away the 3d Ag 500, remove the special designator name of ATLANTIZ and leave this Globe as the LOCK DOWN ATLANTIZ and badge the star as LUU SY FHAAR.

The difficulty with this Galactic Group is that they revisit from time to time, as this is one of their Designed Footstools. It is unfortunate for the local population when the Dogma is so far out of step with the reality, that the Galaxies and the Local Star have to take action, to restore some sort of balance. Yes the Reformation restored some balance, but it was brutal, and is only now accepting that the brutal system does not work. That may be the case locally, but on the streets of the capital, there are marches insisting on brutal regimes with zero tolerance and trampling all over their local populations.

Such have no grasp of the Galactic input, which is marked off by the volcanoes. Yes a major Icelandic volcano may erupt and temporarily disrupt flights, but so what? A minor Icelandic volcano erupts, and still the local Zealots say, "so what?" There is increased Solar activity, and the Locals again say, "so what?"

What I understand other may not and I must not force it upon them. So I have six books written, published by me so that it cannot be said that I did not leave a record, but as for this seventh, why should I bother?

I am satisfied the understanding I have provided with these texts discloses the advanced reality. On Television, I can check on the UFO hunters and the so called Alien abductions, but I regret to say such does not fit the profile for a KHRYZTAAL GALAXIEZ visit. Those who can shape shift to the restricted styles can easily perform the miracles, but why should they again visit? If the RTTUR MNZYVZ sends his official YYG ZYVZ and he is murdered by the locals leaving it to RTTUR MNZYVZ or one of the other RTTUR to restore that murdered official to life, why visit again? As a checking exercise perhaps, but I am aware that RTTUR MNZYVZ will be looking for certain specifics and RTTUR THNGME will be well aware of the on-going problems with the reality; including specifically those thousands of millions who conspired to murder his YYG and who would do so again as a further visit would not conform to the dogmatic requirements of the locals. As then so now; and that is a very serious disappointment to the Galactic RTTUR.

On that basis RTTUR MNZYVZ and the other 23 RTTUR would say the matter is proven, with only RTTUR THNGME holding out with a NOT PROVEN.

RTTUR MNZYVZ points out to RTTUR THNGME that the locals who use NOT PROVEN wish to have it removed to obtain more convictions on sexual matters; something which fundamentally attacks the MILKY WAY style.

RTTUR THNGME points out that although LOCK DOWN has been APPLIED for the 5,101 years going forward there is an 11:150, and that 11 does not expire until 2030.

The other 24 RTTUR defer to RTTUR THNGME in the manner of Revelation 5:14[The 25 is 24 Elder RTTUR plus THE I AM THE!].

≈ 6 ≈

Messenger Guardian Seven

To see face to face!

The photograph has been REDAKTED.

It is acceptable to leave the FRONTISPIECE photo in place since that is the AT FORMAL CERTIFICATION, DULY AUTHORISED.

The plates explanation of the different Galaxies has been provided, including one with the £2.00 Britannia representing a very ancient and advanced FOOTSTOOL. That makes the point that it is way above the Minimum of 10/6d, and with its date it represents new money.

Effectively a new understanding!

While the ancient Egyptians might have shown a vulture in their paintings with the NYTYR, in the modern understanding a "Spruce Goose", might make the point, although it might be said to be an eight winged creature.

The £2.00 coin makes the point about an extraordinarily advanced civilisation, and one of the travel systems used is indeed the Boat of Millions of Years. However that is only one of the systems.

Given the vast distance involved, they can also do "Hosting". That means that they put their understanding into a local individual and thus run the YYG, [Their system Guardian] in an ordinary local. The local will have limited knowledge of the technical aspects of the Galaxies, but using that individual they can check out the local footstool. This would usually be done in conjunction with the Local LANAZT star, as the star would be required to give the necessary assurances of safe passage and residence.

It is the Star request to be readmitted to THE KHRYZTAAL GALAXIEZ, and thus has to request the YYG. In Revelation[22:20] John asks for the return of the YYG[The Lord KHRYZT YYG ZYVZ], completely unaware of what he is asking for. THE GREAT COUNCIL, the TWENTY FIVE RTTUR would require to agree, and the YYG they would likely send would not be the same as the previous official.

In the first place MNZYVZ would not agree to send one of his GYDRAN officials, after the mess made by Pontius Pilate. That would mean one of the others would have to do so, and there would be no point of sending anyone other than the YYG THNGME.

It was MNZYVZ and THNGME who were in discussion with the YYG at LANAZT. The local LANAZT star could listen in to the chat, but both THNGME and MNZYVZ had spotted the old, old problem. The lack of free speech, with the insistence of murder for those who disagreed with the local dogma!

It is not altogether easy to spot in the New Testament, but if it is known what to look for it is there in various places.

Matthew 15

The reference is to dogmatic "tradition", and pointing out that the tradition is not that of YYG ZYVZ Verses 2 and 3 and Verse 9 makes the point that the local tradition is "of men". Fundamentally since it enshrines the dogmas of murder and mutilation; it is NOT of HEAVEN.

Both MNZYVZ and THNGME point out to the LANAZT STAR that on this point their YYG currently in place is entirely correct. That placement has been there for three score and ten, albeit technically that does not expire until the late in December 2024; to make the insistence on 70 years completed.

Meantime both will check on safe civilian air travel; and will continue to do so, on the basis that 3d Ag 500 continues to apply as the Galactic assessment.

The 2/6d and above civilisations, well able to do many miracles using their technology and understanding, are very difficult about who may use it. Generally they will only permit of their own individuals and to their own style. They take exception to others trying to steal their technology. To try and back engineer their technology they would regard as theft; and thus would withdraw their technology.

Antibiotics are their technology, and these are not being upgraded. In respect of certain very ancient and advanced antibiotics the restricted licence is withdrawn INSISTED.

Both MNZYVZ and THNGME are aware of major difficulties in allowing their TZUTRAN GYDRAN to remain on the Footstool.

The local Legal system does not meet their PROTOKOLZ, and thus the RTTUR are entitled to insist on immediate recall, and to LOCK DOWN. On the representations from their GYDRAN such has been delayed, but only to the extent that the 11 year grace period from 2019 has to run.

The cancellation of individuals who insist on correct science over dogma is not acceptable in Galaxy terms, and on that basis **current military conflicts reflect the Galactic insistence on clean up.**

Such tidy up will require full disclosure of various aircraft incidents where craft were lost. Both MNZYVZ and THNGME are highlighting specific incidents, where the locals have done grossly offensive cover ups. THE RESEVE AND REDAKT instruction is given to their GUARDIAN.

The Guardian has asked for exhibition of a "Boat of Millions of Years" He is aware that it is massive, and would give the appearance of defying the law of gravity, except of course that it uses gravity wormholes to do what some might regard is instantaneous travel; a bit like travelling through a door in space and time which is exactly what it is.

Super advanced technology which is shared with those at 2/6d and above Galactic weighting!

≈ 7 ≈

Half Crown to Crown Plus

THNGME has very severe restrictions on what it will allow to be said and seen. It is indeed censorship, but it is essential.

Most of THNGME Galaxies are very liberal, and insist on freedom of expression. Yes there are limits, but they will not allow in anything which is hurtful or harmful, and that specifically includes vicious Dogmas.

THNGME draws to the attention of the local LANAZT star that their Guardian has serious concerns, not only on legal aspects, but on freedom of speech, correct medicine, which includes biology and treatments. THNGME points out that he will not lift the block on medical advances and again reminds the LANAZT star that mutilation is NOT ACCEPTABLE. That includes extraction of teeth. It includes the practice of circumcision, and that they will not allow into their Galaxies those circumcised. It is a matter not up for discussion. LLYKOTT_LLYDRAN may take those circumcised, but for the most part The KRYZTAAL GALAXIEZ will not take such. It should not be asked.

These are comments made by those civilisations on globes assessed at 2/6d plus. This is to do with matter transfer technology; which can be expressed as enabling the deaf to hear, the blind to see and the lame to walk. So advanced is that technology, that in the main they will only communicate with those civilisations weighting 2/6d and above. Those at 5/- plus have had that technology for thousands and millions of years.

One of the difficulties is that many of their cities might resemble the suburbs of a modern technically advanced civilisation, from the US, Europe, China, Australia and New Zealand. Some of these, despite being millions of light years away, have the technology to tap into such as the local internet here and to obtain such information as they wish. Most operate a department Guardian Seven who collect and assess the relevant Globes and they are checking for Compatibility.

This Location ATLANTIZ is a Footstool, for THUURRNN Galaxy, and at this location they have allowed a considerable variation from what is acceptable, such that large areas are REFUSED and only a few are RESTRICTED RESERVED. Thus if any of the 2/6d civilisations wish to visit here, there are severe restrictions as to what is permitted.

It is perfectly permissible for Alien civilisations which have superlight travel, but lack matter transfer, to visit. There are however rather large KEEP OF THE GRASS signs, in the form of various stylised Pyramids which make the point that this Globe is a secured FOOTSTOOL in the PYKAAZ_PYNWHAAL style.

While the local population may be open to visits from Alien civilisations, the very advanced have their own restrictions which mean LOCK DOWN is applied.

Frequently the LOCK DOWN is done with the imposition of a set of rules which do not make the Galactic PROTOKOLZ for a KHRYZTAAL GALAXIEZ FOOTSTOOL.

The words RESERVE AND REDAKT are a clear indication that a society is unsafe. Before any Galactic visitor from these advanced groups, 2/6d + will visit, the local Star is asked to provide locations which can be safely visited. That would include areas where there are no snakes, or where such are easily and safely treated. Thus while Australia would fall into the category of being in the necessary group, the presence of so many dangerous creatures, means that it is not recommended as a location to visit. Other areas are prohibited because of local rules and laws.

A flight Glasgow to Auckland is insisted direct. Yes it can be done by transit through other countries, but the technical point for the 2/6d and above is that the society must have the capacity for safe civilian air travel in comfort and at a reasonable price. Those in THRENZ, M31 Galaxy who are at 5/- plus, potentially £2.00 Ag 925 plus do not do cattle class which is how they would view even business class. This is all to do with the level of understanding. So together with MNZYVZ, FVRGHAAL, THNGME, AAGHAADDWR they block communication and restrict technology. They are content that the locals look for alien spacecraft, and only apply restrictions to any of their individuals visiting. Most of these groups are of the TYGER headed styles, by which I mean Tyger style on a large substantially ATLANTYON Style body. The compatibility aspect is material for the superlight travel with the matter transfer in the correct style.

SEX, male and female, is critical, and if it is not done to their styles, the Globe is marked off with the LUU SY FHAAR term. In simple terms, that means that the local populations are not sufficiently advanced or enlightened to be entitled to see the vast Galactic civilisations. Thus in my case THE GREAT COUNCIL restricts what I may say in my texts. It is the hidden Reality.

Revelation 5:14, the Great Council, Twenty four Elders plus TH NG ME, to whom the Twenty four defer.

THNGME makes the point with the black FLYFLOTT cover, so they will not communicate, and will not permit the connection at this time. The locals can build as many rockets as they like, look through as many telescopes as they like, but they are working with the wrong technology and looking in the wrong place for the wrong things.

≈ 8 ≈

INDEED

The DRAGONSPEAK, as I have previously explained works with particular words, sometimes in ATLANTYON and also in a modified English where the DRAGONSPEAK aspect is shown with the block print, the use of the letter "Z" and the numeral "0" instead of an "<O>" as will also be obvious it works with spelling adjustments, which serve to show the level of development.

With the local insistence on the term "Earth" over ATLANTIZ, it can be seen that the 3d Ag 500 assessment is the best that might be given.

Likewise with the refusal to accept biological sex as the defining description, there is a LOCK DOWN imposed. Only those civilisations at 2/6d and above can change sex using the matter transfer equipment, which is also the rotation equipment. To try to change sex using surgery or drugs is a THROW OUT matter for a population totally lacking in correct medical understanding. So this location now enters the LOCK DOWN phase, and that is evidenced by the imposition of sub standard electric propulsion vehicles imposed on the population and with vicious travel restrictions. The political system is, in Galactic Terms, marked as a serious fail. Yes the local populations can do as they like, but in proceeding against the correct Galactic understanding, it is the Galactic Lockdown which is imposed.

Rocket science is fine for weapons, but it is of no use for superlight travel or communication.

Said THE LORD GUARDIAN to THE LADY RTTUR
AAGHAADDWR and THE LADY RTTUR
ZYRAPPHAMA
"Spelling is not my strong suit."
The reply from both is the hugely technical,
"INDEED".

≈ 9 ≈

The Visit.

I am asked to revisit, and very reluctantly I agree. I am given assurances that in the territory to the North of Hadrian's Wall I will find a legal system where the Galactic Protokols are met, and in which I may have confidence.

The representation is by the LANAZT STAR, MR DRAGON ZYR who is well aware of the difficulties.

The LOCK DOWN was imposed by THREE RTTUR and it would thus take three RTTUR to lift it. As of 1952 the LANAZT star had to accept that the PROTOKOLZ were not met, but on a timescale of within 18 years they might be met. In any event the LANAZT star accepted that the MESSENGER [TZUTRAN] would be "protected".

I do not do splitting of hairs. Such comment is offensive. I do not do counting of Messengers on a pinhead; a likewise offensive comment.

The Boat of Millions of years is arranged to do the delivery. It will take around 365 Atlantyon days for the Boat to travel from THRENZ Galaxy, where the KASTDATA meeting had taken place. The delivery is delayed a short time as the position is assessed, and finally the delivery is made. The matter of uplift has been agreed before departure. Three score and ten must be met, complete, and all the other TWENTY FOUR RTTUR will need to mark content to the arrangement.

There are some serious complications. THE LOCK DOWN will need to be lifted, and as previously the locals will resist.

"How would you intend to play it MR DRAGON ZYR?

There is every likelihood I will mark NOT KONTENT INSISTED, particularly if there are contraventions north of Hadrian's Wall! There are many problems on the Horizon; particularly with the Justice system. North of Hadrian's Wall means there are two legal systems under the electron microscope. The assessment is on a United Kingdom. Divide it and the Galactic connection is lost. You are aware MR DRAGON ZYR, of a matter with which the locals are not familiar, that I insisted on the shutdown of the Scottish Parliament in 1707 because it was seriously corrupt; and that the 1707 Union Treaty is a ZALT Treaty, before the locals even knew what a Salt Treaty was. We are talking serious THROW OUT and LOCK DOWN if it should become corrupt in the future, and that includes all Parliaments in the United Kingdom. You are well aware MR DRAGON ZYR that I can agree to extend the United Kingdom influence, provided it is clearly understood as a feet placing exercise; and I will not entertain corruption

So what do I find in 2019 AD 107? What is reported to me by my various RTTUR?

I find human blood sacrifice writ large on the streets of the Capital, with no desire to clean up the mess. I could say both capitals, but MY RTTUR are correct in insisting that we lock down. Given my presence on the FOOTSTOOL they grant a concession of 11:150, but what do I find but more corruption. Serious corruption with no desire to clean up the mess!

I could call for a new Highland Clearance, and you are well aware MR DRAGON ZYR that there has been put in place the means to impose a mass tidy up. THAT is not my style. My style is to reserve and redact and to withdraw. The withdrawal comes with the removal of our styles, technology and medicine.

Murder suicide by pilots means that you do not have safe civilian air travel. Mid air explosions of aircraft with bombs in them is not safe civilian air travel. My Guardian Department Seven have provided me with extensive lists of such incidents, and as with

Titanic, which was a maritime matter, with judicial cover ups, I hate loathe and detest judicial cover ups. They are unacceptable on any globe which aspires to be a member of our group. So I expect, regardless of how uncomfortable it makes the local Governments, full disclosure of such cover ups, and within the 11 year period applied in 2019, if the 5,101 year LOCK DOWN is not to be made an INSISTED matter.

The AD date reference is material. RESERVE AND REDAKT!

I am talking Galaxies here MR DRAGON ZYR, not Constellations, and the reach of our system Galaxies is vast. THRENZ which is nearest to ATLANTIZ_DRAAZ will need to be satisfied on point, as will we THUURRNN; especially our THNGME Galaxy.

So I check again the text of MATTHEW and for the DRAGONSPEAK therein. You and I both know what the "raiment white as snow" reference means, as also the gardener reference, and that we have severely restricted what might be written.

On that point you will be aware MR DRAGON ZYR of various blood sacrifices made because of what was written by some. I would point out that such is entirely unacceptable. I suggest you make it plain DREAMSEQUENCE that has to stop immediately, otherwise NORTHERN PYNWHAAL and SOUTHERN PYNWHAAL can do a Global reposition, which will make their tales of a flood minor by comparison.

I have no intention of starting from the beginning, but I can remove to elsewhere in DRAAZ, or THRENZ or others of the several of our System Galaxies, before doing the global spin. I suppose the locals could call such spin an Apocalypse, but as you well know MR DRAGON ZYR a minor reposition of the Globe is required to link up with the System Galaxies, to put them so to speak "IN PHASE" for the superlight travel.

ZERO I dislike except as a cipher and balance point, between polarities.

With the DRAGON Stick, DRAGON GOWNS and FLYFLOTT, I can move about the City, which would mean, that the increase could be almost up to 1/-, which is the superlight travel without the matter transfer, but it remains at 3d Ag 500, because I am not safe to walk about the City in formal DRAGON GOWNS as the locals have turned to corrupt and Satanic murderous mutilation practices.

I trust I have made the position KHRYZTAAL KLAAR to you MR DRAGON ZYR, and to HUMMINGBIRD."

Dated 4 January 2024 AD 107.

PROTOKOLZ

This chapter reflects a conversation with MR DRAGON ZYR of what RTTUR THNGME requires. There is no indication that the LANAZT star has accepted that, but if it is not done the LUU SY FHAAR designation will be re applied and the 5,101 years with its 11:150 will have to run.

There are various oblique references such as the requirement for the legal systems to be balanced and fair, and not to have harsh and oppressive penalties on the Horizon

Satanic, means female. It does not include female in the sexual sense, but satanic practices would include any mutilation practices with under 18's and would include circumcision, and hormone treatment or gender re assignment. Regardless of what the local laws say, such gender re assignment is viewed in Galactic Terms as a satanic practice on which the use of Nuclear weapons may be authorised.

My texts here set out the Galactic position, NOT the local one. At a Galactic level the locals here are in serious trouble with their "gender speak", "no platforming" and other politically correct ideas and corrupt practices with such at Net Zero. It is for the local Star to determine Climate, not locals here. Thus the comment about repositioning the poles! Given that Amenophis IV had a rough idea of this Science, it is unacceptable that the locals now talk about "man made climate change."

The comments here are for the Local LANAZT STAR on its application to be re admitted to THE KHRYZTAAL GALAXIEZ.

II

Said MR DRAGON ZYR to RTTUR THNGME.

"A little touch of plain speaking LORD RTTUR!"

Said RTTUR THNGME to MR DRAGON ZYR,

"I have revised MR DRAGON ZYR. On the Highland Clearance point you might point out to them that excessive, unfair and vindictive taxation will do the clearance very nicely; a matter of which the locals seem not to grasp."

Said MR DRAGON ZYR to RTTUR THNGME

"Well spotted. I am aware of THE KHRYZTAAL GALAXIEZ PROTOKOLZ."

≈ 10 ≈

Pronunciation, Passcodes and spelling

In all of my previous books an alternative spelling has been used, most usually with technical bits as might be viewed as passcodes.

Thus while books 1-5 dealt with Atlantis, it was only with the number 6 that I changed the spelling to ATLANTIZ, and in blocks. That makes the point THAT the matter is correctly identified as GALACTIC.

Thus in the King James Bible at Page 55 of the Oxford print at 3:14 are the words "I AM THAT I AM". Fundamentally that makes the point that the individual in question has Galactic connections, and in terms of what I have written in conjunction with Revelation 5:14, THAT individual is RTTUR THNGME; who if I might add has many names and he and only HE determines the name used, the number ONE of the TWENTY FIVE RTTUR.

It is all "passcodes" and "pronunciation", and immediately I use the double "PP" there is the reference to PYKAAZ_PYNWHAAL and the specific design requirements for a water based FOOTSTOOL. I sometimes change from the letter "Z" in FOOTSTOOL simply because it looks untidy, but I do use the blocks to make it clear that it is a GLOBE I am talking about, not simply a piece of furniture. And yet, in the scheme of this Galaxy as ATLANTIZ_DRAAZ, it is a piece of very valuable furniture. A piece of furniture on which an RTTUR or KURALAKYZ may place their feet as a checking exercise.

While RTTUR THNGME might have preferred to visit himself, given the mess with Pontius Pilate, such a move was considered unwise at this time, especially as the old LOCK DOWN matter was still in place.[Watch the P P in Pontius Pilate as if, as a reminder, that both PYNWHAAL GALAXIEZ are watching EVENTZ.] There is further annoyance when post 2019, that lock down is insisted in the territory to the North of Hadrian's wall, where it has no right to be in place.

The Enlightenment requires free speech and correct scientific understanding; not dogma. Put dogma in place and buildings will be destroyed; most often by fire.

Ghettos and slums are not acceptable and it falls on the leaders of the States to provide more than adequate housing for their populations. Slums are cleared by "blitzkrieg" only when no other course is possible; particularly where the politicians have dithered and delayed to provide high quality accommodation.

I had anticipated providing an outline of the City of Kyen, but the Lady KYEN TY AMEN TY will not permit. At best she will allow reference to the *Reuymans* factory at North Beach, Kyen, district Reuka where some of the Boats of Millions of Years are manufactured by the *Reuymans Trust*. These Trusts are similar to a plc, except that the factory has to provide adequate accommodation for its workers; a bit like Port Sunlight and Bourneville, and that the workers are all members of the Trust.

I am talking about very advanced civilisations here and they will thus look very closely and carefully at what is on the ground at this FOOTSTOOL, to assess on compatibility.

Reuymans manufactures aircraft for use on its THRENZ footstool, and some are done with the anti gravity vacuum lift. The ones which are Galactic use the magnetic wormhole drives, and it is that advanced technology which is restricted and where other globes have to demonstrate their entitlement to have that technology.

Once an aircraft the size of an A380 has been produced, *Reuymans* and other aircraft manufacturers in the KHRYZTAAL GALAXIEZ take an interest. They can see certain technical points which mean that civilian air travel is far from safe. In the main it works, but an accident on the runway should not leave the arriving or departing craft in flames, with risk to the passengers lives. For that reason much of their vacuum lift craft, although large, take off and land on water, and thus have to have flotation requirement to enable passengers to safely exit. As with such as Sydney harbour injured passengers cannot be allowed to remain in the water because of large aggressive fish.

Where land based runways are used, the hull of the aircraft must be able to sustain a level impact landing at take off speed+, so that specialist materials are used, with shielding equipment. That is sometimes a wrap round bandage, as the craft is built. Many of their craft do not have windows, except for the forward view for the pilots, and internally it is TV screens, or a pixelated screens, sprayed on to the inside of the craft which give the passengers the appearance of windows. When the Pixels are sprayed to the inside of the craft, the view can be done to make it appear there are no sides, roof and floor. Such is not often done as it can make the passengers feel unwell.[Motion sickness with vertigo]. That technology is primarily reserved for the Boats of Millions of Years. Such technology does allow for the inside of the craft to appear as something entirely different, as with the stomach of a Whale, without the fluids.

Such technology has been developed over thousands and millions of years, and is not provided to thugs and murderers. Pilots of such craft are strictly trained, and anyone regarded as unstable would lose their pilots licence.

Yes these advanced cultures have retraining facilities, but given that many of them designed the NYKYDYNYZZ material, which is the design of the species, they have access to very effective reprogramming, including rotation.

Locally as this Globe is weighted at 3d Ag 500, the locals do not have reprogramming or rotation facilities and the so called correctional facilities do not meet Galactic Criteria; thus much technical and medical information is withheld. They need to progress from the 3d Ag Globe to 6d, 9d and 1/- plus.

It is fine for the Science fiction to be produced, but much of it lacks technical accuracy. If the NYTYR of Egypt continue to be shown as "the baddies"; that discloses a lack of correct understanding.

That lack of understanding is not always obvious.

To refuse to carry out a blood sacrifice or judicial murder, discloses a satisfactory level of enlightenment such that Abra (not enlightened) may be changed to "AbraHAm", with the letters "HA" representing the enlightenment; the Galactic ZY FHAAR. But where the enlightenment is lost, as when the mob demand a blood sacrifice, then the crowd is lost and the LUU SY FHAAR term is applied. The population are not enlightened.

A peaceful demonstration on the streets of London may be said to be "enlightened" and does not cause a problem, but when such descends into calls for murder, only the term LUU SY FHAAR is appropriate. Thus that matter is recorded at the trial before Pilate with the locals, including Pilate being unaware of the Galactic implications.

Rome is built within the caldera of a volcano (the seven hills) and it is thus only safe provided the Galactic RTTUR do not authorise volcanic activity. Vesuvius is the vent to take the pressure off Rome; and that is vulcanology and Galactic Science.

As I explained previously, those volcanoes which are inlet valves for Star Plasma sent to this Footstool, are HYHASTERN, those which return the plasma are KULYN. When the KULYN go active that is usually because there is a request for recall of plasma as locally something wrong has been done.

DO NOT ATTACK the WHALES. They are designators of the PYNWHAAL GROUPS, NORTHERN and SOUTHERN.

Mr WILD 0RKA ZYR is a moniker for both PYNWHAAL groups.

The pass codes are simple and complicated. PYNWHAAL is a correct spelling at an advanced level in Galactic terms; likewise ATLANTIZ_DRAAZ. Different spelling may be used, but like a pass code if the wrong entry is made the LOCK OUT or LOCK DOWN continues to apply.

My ATLANTIZ texts may appear as fiction, but they are pass codes. As such to change or alter even one word is likely to mean the pass code will not work. Equally difficult is that the passcodes change from time to time.

MERLIN might work, but drop a letter or two and you end up with MIRIN. That can be changed to MRLY2N, and equally to MRLYYN.

THNGME is another and a very restricted name. TH NG ME, which highlights that it is THUURRNN and a TRINITY aspect.

In this chapter when I am dealing with passcodes it has to be understood what each variant represents, and at what time and place each may be used.

The text is submitted to THE LORD THNGME for approval, or for RESTRICTED RESERVED to be APPLIED.

≈ 11 ≈

Visions and Visitors

Details of the alien visitors come from all over the globe. Mostly they are dressed up as philosophies, and occasionally some have visions. My attention has been drawn to Tesla, now a name common with electric cars, and what may be said to be the visions of Mr Tesla. When he suggested that the information he had obtained, might have come from Mars he was ridiculed, and of course reference to "little green men" and Aliens is a source of ridicule today. I can easily change Mars to MA R Z; which is to make a Galactic point. The ONE RA GALACTIC, for which I use the Galactic Term RTTUR. I could also say RTTVVR; but in all instances the sound has to come out as Arthur.

MA R Z, instead of Mars, thus discloses an entirely different understanding, since the use of "Z" means it is Galactic, and from the Atlantyon, "MA" is one. R of course relates to the local star. So in the case of Mr Tesla, his "vision" did not come from Mars, the planet, but from the local Star, which as Amenophis IV realised provides for all life on this Footstool! The local Star is a life form, but not appreciated locally here as such.

The local Star is also the means by which material may be delivered from different stars in different Galaxies, and thus also the means of communication star to star. This information may arrive as visions, and is likely only to be provided to those who have a clear understanding of what they are dealing with.

Viral treatments can be delivered by Star visions.

Some of the old myths convey the information, as with Prometheus, but in that form will tend to hide far more than they disclose.

The style of the visions can vary, and in this time frame, the material can be sent via the television and on the Internet.

I watch a number of films on UFO's and as far as I am concerned, they are all fake. I could say deep fake. That applies also to stories of Area 51 in the USA. However in respect of Area 51 there is a quite different interpretation. What if the comments about area 51 are to direct attention to the Whirlpool Galaxy, M51 and the RTTUR FVRRAKOTHERA?

Certainly the large city there of FVHAALYA, with its KAZTDATA building sitting high on the hill, might resemble Mount Olympus, where the Gods meet. That might also apply to the description of Ben Nevis in Scotland, but there is no large meeting hall there for 25 RTTUR to meet. Besides the KAZTDATA for ATLANTIZ_DRAAZ, is in the ATLANTYON ZYON on DRAAZ, some 15,000 light years away in this Galaxy.

Given that these advanced Galaxies I am describing have access to the local internet, why can they not send some photos of that Atlanton ZYON. Indeed all those cities in THE KHRYZTAAL GALAXIES, would refer to them as ZYON.

Part of the explanation is that for many on this Globe ZYON is a dirty word, and these powerful Galaxies take considerable offense at that view. Why should they give any of their technology to such ingrates?

This Globe is a restricted secured THUURRNN footstool, and even at 3d Ag 500 there remain considerable restrictions.

I am restricted from commenting, even correctly on certain matters, because politically correct means that only certain dogmatic views might be currently expressed. The Global population, in Galactic terms must thus be viewed as unenlightened, or indeed as Satanic. The term Satanic, applies to specific mutilation practises, and even light BDSM would fall into that category. Satanic in

Galactic terms is female, and the application of the term is to a wicked woman imposing on a weak husband. Lady Macbeth is an example.

There is a particular Galactic understanding. Sex is not only permitted, it is encouraged, but it MUST be safe. Thus online material which does not meet the not under 18 PROTOKOL, would be viewed as Satanic, as would any attempt to change sex. Gender identity causes a far greater problem than is apparent, and for such bad behaviour the Galaxies can recall such as Antibiotic treatments, and refuse to provide the up-rated design material or advances in medicine. All effectively unseen matters!

Puberty changes are STAR CONTROLLED, simply by walking out in the daylight, and thus so called "puberty blockers" are seriously wrong medicine. Any doctors who perform such are not performing correct medicine, and have contravened the Hippocratic Oath.

I have to deal with certain old matters.

The Ananuki, for example, does not look Galactic, but is. In this text it is read, left to right, but I am aware that certain languages read right to left. Thus I have to change Ananuki to KY ANANN; which looks suspiciously like THUURRNN as written, and would relate to THAT male, He of Exodus 3:14. The Head Visible Invisible, of THE TWENTY FIVE as I might refer to him.

So I collect another piece of the Galactic Jig saw, and I trust correctly put it in place.

≈ 12 ≈

Sun Cycles.

Mr DRAGON ZYR and HUMMINGBIRD also discloses an advanced understanding; a Dragon matter.

The Galactic connection between the local Star and the Galaxy should be obvious, but up close and very personal it is difficult to see. An Aid memoir may be given with various "God" figures, but they are more than likely adulterated by local perceptions and dogma.

The canvas of the Galaxies is enormous and it is easy for the locals to make mistakes and to misunderstand. Thus my previous books all contain errors of various kinds, but sometimes these errors are essential. It keeps those who are unwelcome away from the Galactic mechanisms, which can be dangerous. Thus the DRAGON term!

The MR DRAGON ZYR term is deferential and polite, and so also is HUMMINGBIRD.

The local star HOSTS individuals and this HOSTING aspect is typified by Baptism. The Ancient Egyptian headdress with the cobra highlights that HOSTING can be exceedingly dangerous, and it is a matter which is in the Hands of MR DRAGON ZYR and HUMMINGBIRD.

To travel to distant stars the internal brain thoughts MUST be compatible, both at departing star and receiving star. If they are not compatible no travel; and the gates remain closed. The local star can rotate an individual, but cannot send to another star. Certain of the mechanisms are instantaneous, as if safely walking through a doorway. Where a new body is provided it will most usually be in the form of a new born child. This INVISIBLE aspect is a security nechanism

Sometimes, with superlight travel, mechanical equipment is used as with spacecraft, of the kind which uses the wormhole travel. Such are the 2/6d and above. Above the £2 coins, as with Britannia, the civilisations are likely to do instantaneous travel from an individual living to another living individual. Given such transfers and rotations the span of years can extend to thousands and millions of years, and numerous styles, the bi pedal Atlantyon, Tyger, Wolf and Rhino.

In this Group those four styles are INSISTED.

≈ 13 ≈

Puzzle Pieces

To make up a jig saw puzzle it is helpful, if not essential that the design of the completed puzzle is known. It is also helpful to know the number of correct pieces, so as to exclude those from another puzzle which is not in the picture.

The global puzzle, is complex, because the correct pieces have been scattered all over the globe at different times and with different understanding. It is all one to say there were alien visitors, but where had they come from?

That question takes on a new meaning wearing the fancy Dragon Gown, with FLYFL0TT, if asked where I come from I say Glasgow. That is both accurate and misleading, because the gown and FLYFL0TT are not items known here. The pointed FLYFLOTT looks like a serpents tongue, so if I was to be asked if I am KWTLZ KO ATLZ, I would reply with a negative.

My last book displayed the Dragon Gown and FLYFL0TT , but the FLYFLOTT was unclear, except that it was not black. Thus an open Galaxy, but in this book the FLYFL0TT is clearly black and very pointed. The last book said ARRAN K0 ATLZ, but if I am asked the same destination question I again say Glasgow. ARRAN K0 ATLZ is not known here, so to the locals it is meaningless.

To a certain extent THAT was the problem Moses had with THAT tart reply recorded at Exodus 3:14. The individual knows exactly who he is and from whence he has come, and why he is visiting, but that is not information to be conveyed to Moses. Some thousands of years later THAT same individual is doing similar checking, albeit slightly differently, but the checking process has produced the same result. To a certain extent it usually does.

I check on Gilgamesh, and on the script. I am not interested in the local interpretation, but on the ATLANTY0N. I do not see Gilgamesh in the written text, but I do see the Atlantyon letter "g". I am not sure if it is lower or upper case, or if they even had such. The letter "G" highlights, as does the MAYAN script that there is a Galactic connection. In the case of the MAYA, in Atlantyon script, the letters of the word are split as an aid to pronunciation, and ciphers are added to designate Gender. I should really say "sex", because it is the sex which is relevant, not the artificial construct of gender.[GG LL 3 AZH] That looks to be LLYKURRYON, M64 Galaxy, where the AZH is for Galactic Enlightenment. AH00RA_MAZDA is similarly written in Galactic terms as a double Galaxy and Enlightened group.

Gilgamesh is connected with THE KHRYZTAAL GALAXIEZ so I needed to change the spelling, and the word order as with YYGDRAAZAL, because there will be a concealment aspect with the name. That applies to the individual. The pronunciation is similar, but the Galaxy interest is only with correct identification.

It is similar with Ananuki which in the previous chapter I had changed to KY ANANN, but I had not dealt with the letter "U". THAT was actually staring me in the face, as "00"; the double Galaxy marker, because KY ANANN 00 KY, double "K" double "00" is none other than THNGME, HEAD RTTUR of THE KHRYZTAAL GALAXIEZ. And he is visiting to check out one of HIS designs. THE NYKYDYNYZZ material is his, the so called "dna" is dead material, it is not the living individual!

I am talking here about 2/6d and above footstools who can remake individuals using the matter transfer and transiting equipment.

So what I now have is the visiting RTTUR KYDRAN checking out the level of development on one of his FOOTSTOOLS; checking in particular if his design specifications have been met.

And the answer at certain locations and at various times is in the negative. Thus it will appear as a plethora of different Gods, but it is invariably the same individual, who says simply I AM THAT I AM, and when reference is made to his officers, the usual reply from his officers would be THNGME, or perhaps slipped in as UR and THIMMINIM. 0f UR; or more correctly, from that KHRYZTAAL GALAXIEZ GROUP [00]: from the stars.

As for the local understanding, it is light years out. Not an extra planet in this solar system, not even Mars, not a constellation either but a far distant Galaxy Group, known as THE KHRYZTAAL GALAXIEZ, where the nearest one to ATLANTIZ DRAAZ is M31 THRENZ.

TTZAL, could be pronounced as Tesla, and Nicola could be written NY KO LA, where the masculine is ending AL, where the female would be LA , and so that masculine and feminine gender is transferred to names; as is currently common.

Sex is of specific importance in the PYKAAZ_PYNWHAAL group of Galaxies.

The "Z" as I have explained is a Galactic reference; provided it is correctly understood as such. Change the Z to an S, and the Galactic connection is lost.

Currently the powerful countries build rockets, but such is seriously wrong science for Galactic Travel. Thus certain of Mr Tesla's designs could not be used as they would have interfered with the existing "wormhole" mechanisms, to which access is not currently allowed.

The LOCK DOWN mechanisms vary, but fundamentally they involve repositioning the globe. Not by much, just little tweaks. I say little tweaks, but they will involve melting and reforming the polar and anti polar ice caps, and thus the magnetic poles move to make the necessary adjustments, so that there is correct spin speed. The gates are only open at certain times, like a causeway; a bit like the 29 February. The control mechanism is the local star; a life form, but not in the style currently appreciated.

It is the local Star which provides the sunlight, which is part of the mechanism, but in the style MR DRAGON ZYR, or as HUMMINGBIRD, and it is that mechanism which does the styling of individuals at puberty in particular. The obvious male and female temperature selection is done by the egg laying species, some of which are the modern dinosaurs. For the live bearers, as with the Atlantyon population it is done at birth when the child is safely delivered and the Galactic input can be put in place; adjusted at puberty and recalled when the allotted time is up; extension requests considered but not always given.

The THUURRNN Galaxy NGC 1300 is rather far away, and it uses the strict criteria that "nothing which is hurtful or harmful is admitted to their "GALAXIEZ." In short form their GALAXIEZ can be referred to as ZY 0NN; with the double NN signifying THUURRNN connection. The letters TH also indicate THUURRNN connection; thus TH 0 MA Z. The zero inserted to make the Galactic point, and it is to TH 0 MA Z that the task fall to check the piercing.

"One does not suspect, Lord THNGME. You know full well."

MR DRAGON ZYR makes the point that HE checks all texts, particularly when he is in MR DRAGON ZYR form. He is well aware of my mumping about the gender nonsense, and of the corruption in various of the Global legal systems, and of course Net Zero which does not compute, is not Galactic and where the letter "o" is entirely appropriate and the "0" is not.

"These are local problems, with which I am working" is the rather Tart reply, for I am well aware of the "work in progress" like this text. I complain that this text is difficult, and the response is from MR DRAGON ZYR.

"Simply work with it. I will guide as I have always done and will always do."

The chapter is ended. Enough clarification has been given.

TE ARIKI KAHA RAWA

≈ 14 ≈

Dragon Tidy Up.

Mr DRAGON ZYR and HUMMINGBIRD have various options for tidy, and for checking.

During what is a HUMMINGBIRD period the local star will appear to run smoothly! That does not mean there will be no problems, but they are likely not to be appreciated for what they are. The volcanic eruptions of Tambora and Krakatao occur within different star phases. The 1815 Tambora eruption is in the MR DRAGON ZYR phase, while the Krakatao eruption in 1883 is in the HUMMINGBIRD period.

It has been suggested in certain quarters that the heavy rains caused by the Tambora eruption contributed to the defeat of Napoleon, and certainly after Waterloo there was a change. As to the DRAGON involvement, if it was the heavy rain which caused the defeat, and as rain is H20; a trinity symbol, then it can fairly be said it was all the handiwork of Mr DRAGON ZYR; and that would include the late arrival of the Prussian Troops. As for the banishment to St Helena, given that that is a large Atlantic volcano, then that also is Dragon handiwork.

The usual expression, masked for THAT DRAG0N is that "the Lord works in mysterious ways his wonders to perform." I am not talking about sun worship, but with the Galactic Science, and the understanding of the critical importance of the local Star. Care has to be taken not to overstep the local Star nor to tread on his DRAGON TOES nor to irritate HUMMINGBIRD.

The volcanoes are all controlled by the local star, but various Galaxies supply plasma to the Local star, some of which is stored underground at this FOOTSTOOL. Where there are problems those distant stars may ask for return of their plasma, regardless of local difficulties. Thus FUKUSHIMA where both KIZMEN and SAKURAJIMA were involved! Those stars requesting the return of Plasma like Galaxies Whirlpool and Northern Pynwhaal. Do not hunt the whales, nor the sharks! If you do that it is likely to provoke those Galaxies who are likely to insist the local star takes action. The local population are oblivious as to what is causing the problem.

The Fukushima incident occurred during the HUMMINGBIRD phase, so was likely far smaller than if the matter complained of had occurred during the MR DRAG0N ZYR phase. The story of Moby Dick and the white whale is a Galactic reference to Northern Pynwhaal Galaxy, where in the Northern Atlantic and Arctic are to be found the white whales. In DRAG0NSPEAK terms, as indicated in Alice in Wonderland to a certain extent words mean what MR DRAG0N ZYR and HUMMINGBIRD wish them to be. And the mask stays firmly in place so that MR DRAG0N ZYR is a seen and unseen!

So the masked figure in the photograph in this text represents the Galactic Dragon, but that is only if the individual is H0STED. Otherwise the figure is simply an actor.

The star phase now is MR DRAG0N ZYR and a tidy up is required. What that tidy up will be can be difficult to see. Certainly the local politicians and the mega wealthy are loading on to the backs of the poor a load the poor cannot be expected to bear, and their cries go out to MR DRAG0N ZYR who carefully monitors.

In the DRAG0N phase MR DRAG0N ZYR can take his time, but the checking is most thorough. How are the local Legal systems performing?

Badly does not come close! Exceedingly badly might be more accurate. Various territories have upset the DRAG0N in the past and action has been taken. Thus what was "The Middle Kingdom" has now been relegated to "The Middle Kingdom that was" as the Mandate of Heaven is withdrawn.

In certain respects the Emperor is an individual mandated by Heaven, but it becomes difficult when that Mandate is abused. It would have been sufficient for Cromwell to have accepted the removal of the King, but the blood sacrifice insisted upon, was NOT conform to Galactic Specifications(a matter which ought to have been appreciated by those conversant with the Christian tradition) and thus the Mandate is removed from Great Britain. The insistence on removal is by THE GREAT COUNCIL and by THNGME in particular. For the Mandate to be restored to the United Kingdom, the Kingdom must remain united, and be a ZY ON location and it is THNGME who must restore it. It is Samuel who is entitled to place the ATTYFF on the head. So both HUMMINGBIRD and MR DRAGON ZYR make the point that even with Charles II, the MANDATE was not restored with the restoration and it is DRAGON EYES which watch very carefully during the reign of KARL0S III.

METHINKS that a certain Archbishop …

"You Do Not METHINK, L0RD GUARDIAN. You know!"

MR DRAGON ZYR makes the point. Galactic chess is very real, and it is necessary to know the pieces. The King is not the most powerful piece on the board.

RESERVE AND REDAKT.

MR DRAGON ZYR is aware that THNGME is becoming very difficult.

Yes he has agreed to the 3d Ag 500 weighting, but is not impressed by the difficulties with civilian aircraft, which MUST be safe to move to the upgrade.

As for the legal system, that does not meet the THNGME protocols. Indeed the other Twenty Four likewise confirm that they are not satisfied. Part of the difficulty is that they regard a suicide as a blood sacrifice, and THAT is not acceptable.

Are the streets of the Capitals, North and South safe? The answer to that is in the negative, which makes the whole matter NOT KONTENT, in Galactic terms.

It is thus for MR DRAGON ZYR to take action.

The volcano EYE of a Jackal, made the point, but the effect was far too small. PP KHAAT Y PTAL could certainly make the point, and does on a daily basis to the city nearby. Which galaxies are involved with that? That would again be Whirlpool, with its head RTTUR, FVRAKOTHERA, and M64, The Black Eye Galaxy, or known by its Galaxy Moniker, THE BLACK ICE, and its extraordinarily difficult RTTUR, LLOMOTTOR. He does use white Tyger Shape, but he does rather enjoy playing with miscreants in his RED RHINO shape. Red Rhino in Atlantyon shape might be difficult to see, but then the obvious often obscures the reality.

So perhaps as one of the chapters 21-25 I will have to insert a chat between LLOMOTTOR and THNGME, probably the two most difficult in THE KHRYZTAAL GALAXIEZ system. The matter for their discussion? The necessary Galactic Tidy up at the Footstool at the LANAZT star.

Did someone say WW3?

≈ 15 ≈

AT 0N

In my various books I have made reference to the ATLANTYON language. I am not very good at Languages and thus I do not have a complete vocabulary. Occasionally my attention is drawn to specific words and I thus use them. The different spellings reflect both Galactic interest and local interest. As will be seen from earlier chapters here I have to move the letters around to ascertain the Galactic position. I am aware that some reading this may say that it is all nonsense and not accurate, and that may be fair comment. However in writing these texts the desire was to try to understand the Galactic position, and to make clear certain matters which may have been obscured; often quite deliberately by MR DRAGON ZYR and HUMMINGBIRD.

One such is the designation of a FOOTSTOOL in the GALACTIC System, where the AT is for Footstool, the 0N is for the spiral City which is the Galaxy. In certain respects the letters can be moved around to be read differently, but AT 0N is a Galactic matter.

I could be very difficult and say that ON is NOT Galactic, while 0N is, the difference being that the numeral Zero is intended to represent the Galaxy Disc, while "o" is simply a letter. But given the clear DRAGONSPEAK, I suspect that both HUMMINGBIRD and MR DRAGON ZYR have been making the technical point for some time, given that the AT LAN TYK ocean is the star ocean from which ATLANTIZ takes its name. YY R TH, also draws to attention the Galactic connection, with the YY pronounced "E" and the "TH" is short for THUURRNN; YYGDRAAZAL, a similar word shift for ATLANTIZ_DRAAZ, which is the name for this Galaxy when IN SYSTEM.

"Do I have to remind you L0RD Guardian, that where you wish to use the METHINKZ term; you KNOW and should not use the word "suspect."

MR DRAGON ZYR making a very technical point!

The little matter which the locals appear to have difficulty understanding is that THE TRUTH is the REALITY and it DOES NOT CHANGE. Yes it can be concealed (as with sweeping under the carpet) but the Truth remains.

Where the Truth is clearly understood, it can be very difficult to persuade those who are dogmatic, that their understanding is incorrect. In certain respects the dogmatic do not wish the correction and rectification, and are best left to continue with their error; regardless of the fact that it may lead to disaster.

So I review the AT 0N position. This location is a THUURRNN outpost within ATLANTIZ_DRAAZ Galaxy. The locals may not accept that, but that is the position. As such matters here are for THNGME RTTUR, who having considered the position with MR DRAGON ZYR and HUMMINGBIRD is KONTENT to leave matters with them. It is pointed out that Hadrian's Wall is a Galactic Boundary marker with MNZYVZ and THUURRNN interests. The no English rule has a particular meaning and THAT RTTUR THNGME is of a mind to snap a DRAGONSTICK to express displeasure that English Rule has again been introduced, which term "English Rule," has really little to do with England!

English Rule in that regard is a DRAGONSPEAK concept which relates to harsh and oppressive penalties. NOT in THUURRNN TERRITORY.

The RTTUR THUURRNN THNGME does not wish an Earth Crust displacement, which would be an Apocalyptic flood, but wishes matters tidied up before 2030 AD 107. The preference is that WW3 is to be avoided, including by preventive measures. That could include volcanoes.

The effect of that is that MR DRAGON ZYR has been given the activation codes, for not only the volcanoes, but also the DRAGON'S TEETH. While the locals may deliver weapons, it is a DRAGON matter as to deployment and use.

The little matter of Daniel 12:12; and the Activators.

≈ 16 ≈

Galactic Activators.

This is the exceedingly difficult bit. MR DRAGON ZYR is fully conversant with the Activators, but I am restricted in what I might say.

As a UK Prime Minister once said, it is all about, "events!"

When the event is described as a Pearl Harbour event, that is an event which upsets the FOOTSTOOL application for re admission to THE KHRYZTAAL GALAXIEZ SYSTEM, and there may be a SLAP DOWN after 1335 days, and it will often affect two cities; including a "Holy City." For the City cannot be classed as holy, if the locals there worship blood sacrifice! Such individuals are NOT enlightened, and the HA is removed.

I AM INSTRUCTED TO DO A RESERVE AND REDAKT INSISTED.

There is FURIOUS ANGER AT GALACTIC Level.

That is a matter for the RTTUR, but it serves to highlight the difficulty in identifying the activators.

In essence it is for the TWENTY FIVE RTTUR to mark NOT KONTENT, and once all TWENTY FIVE AGREE, MR DRAGON ZYR can let rip.

The discussion between MR DRAGON ZYR and THNGME is thus of paramount importance. Another chapter in the 21- 25?

METHINKZ NOT

MNZYVZ points out that the lack of free speech means the KURALAKYZ is threatened; and that is a star matter requiring immediate Action. The 1335 days have expired. IMMEDIATE action is now INSISTED.

≈ 17 ≈

GALACTIC SLAP DOWN

The Slap down is not seen as such, because it was done locally. But the authorisation for the Slap down had to come from far away, and that included providing the technology for the Slap Down.

Nuclear weapons are, as should be obvious Dragon's teeth, and as such the Dragon, MR DRAGON ZYR as the local star determines who and where they are manufactured and how they are stored.

Some are stored at Military facilities, but others are stored at so called nuclear facilities, where they are not so obviously seen as Dragon's teeth. But when there is a problem, as with Three Mile Island, Chernobyl and Fukushima, it is MR DRAGON ZYR who has intervened, usually to make a point, and not always seen correctly.

In the case of Three Mile Island, the point is made that when containing any Dragon equipment, considerable care has to be taken and tolerance can be very minor and must be adhered to.

A similar point at Chernobyl, but an additional point, which is to beware of what you have stored in your own back yard.

The Fukushima incident is more complicated because it is outright Dragon activation as the THUURRNN and WHIRLPOOL Galaxies make the point that they wish their plasma stored underground, returned immediately, and that regardless of the damage caused. The volcanoes Sakurajima and Kizmen are activated and the resultant Tsunami sets off the Fukushima plant. As to the reason? Stop the whaling!

The Science fiction film had made the point about the whales, but the locals had not grasped the point about the powerful Galaxies who are offended. They do not need to travel here. Their mechanisms are already here in the style of a Local Fire Disc, and a specifically designed FOOTSTOOL; on which from time to time the RTTUR may SAFELY place their feet.

The safety aspect restricts very severely where the RTTUR are willing to visit; and in respect of such visit they must be satisfied on safe medical and legal practices.

≈ 18 ≈

The Enlightened Understanding

In dealing with ATLANTIZ the star point is Plato's description of Atlantis and the discussion about the Egyptian understanding. In essence, with his rough drawing Plato made Atlantis up. Except he did not!

ATLANTIZ, the REAL ATLANTIZ, was the FOOTSTOOL on which he was standing, and the Local star was using him as a vehicle to help to restore ATLANTIZ with its Galactic connections.

My No 6 book was entitled THE LOST ATLANTIZ, with the spelling changed to reflect what I understood to be the correct understanding of ATLANTIZ as being a Galactic Location.

ATLANTIZ in Galactic terms is still "Lost" as these vast Galaxies will not lift the Lock down. THNGME is in VERY SERIOUS KREAM PUFF. He simply cannot believe what the locals are doing on his RESERVED TERRITORY.

I am well aware that THE GREAT COUNCIL, the TWENTY FIVE RTTUR can be very difficult about what information they will agree to release.

For fiction it does not matter what is said, but when dealing with the reality it does matter. I may say FVREKKRAN, and write it as Fvreckran or indeed as Vekran, but the pronunciation and spelling are important as they are code and set the level of understanding and enlightenment. Various Global languages write their languages differently, and that in respect of very large and small groups.

I write in English, and with the revised spellings I am making the point, that there is Galactic information in these texts. It is not forced information, it is simply made available. It may assist some, others might simply bin it as being fiction. The relevant matter is that The Truth does not change. It may be concealed, including as swept under a carpet, but the Truth remains.

The very powerful Galaxies and their Civilisations as I have attempted to describe them are very real. They have left their calling cards in the various locations as pieces of the Galactic jig saw, which I have picked up, and hopefully correctly pieced together.

In their terms this Globe as ATLANTIZ is at 3d Ag 500, and these advanced civilisations and their PEARL HARB0UR are at 2/6d and above; and have been around for a very long time; thousands if not millions of years prior to this current time frame. Some around in a very advanced state at the time of the dinosaurs!

Plato in his text talked of a circular city linked by canals.

MR DRAGON ZYR or HUMMINGBIRD allowed Plato to write, and the canals talked of by Plato, can very easily describe the wormhole or indeed the WHIRLP00L connection. The circular city is THIS GALAXY as a ZY0N. I call this Galaxy ATLANTIZ_DRAAZ, and it can also be referred to as "The Golden Fleece" or "The burning bush" and " the fire that burns and burns and does not consume".

Canals are made, and to a specific design, and so the canals go from Footstool to Footstool; effectively from Pearl Harbour to Pearl Harbour.

That does not mean that Pearl Harbor is a wormhole point, but rather that Pearl Harbour is a term which reflects what this local LANAZT star wishes, which is to be re admitted to the Galactic System

THAT the TWENTY FIVE RTTUR had to agree, and as I indicated previously the HEAD RTTUR THNGME was having none of that. He is NOT KONTENT with conditions on THIZ FOOTSTOOL.

Two criteria on which he has to be satisfied, and where he is not are,1 The medical practices and, 2 an enlightened Legal System, which has to be balanced and fair, and not indulge in harsh and oppressive penalties (description which can be monikered off as "English Rule". It might also be marked off as "the Rule of Law" or "Harsh and oppressive Dogma.")

THNGME Eyes FLASH at the merest suggestion by some that Pontius Pilate was trying to release YYG ZYVZ.

The particular beef which these advanced civilisations have is that they do not wish their very advanced technology and medicine to be misused.

Currently with the advent of digital technology and artificial intelligence there is massive scope locally for misuse, from the misue of digital parking charges to dish out vindictive penalties for minor infractions of timed parking, to storage of dna information for the state to misuse with facial recognition. When is a face match not a match; as with twins as one example.

In terms of my texts I am talking not of dna, which is dead information, but rather the NYKYDYNYZZ material which is the life force and which must be correctly understood, by those doing the super light travel with transiting.

Thus the Aliens must come in particular approved head designs, and body designs, male and female, which is the "milky way design", the general insistence of these advanced Galaxies is "in the old fashioned way."

The DRAGONZ as I refer to the RTTUR, heads of Galaxy, rather delight in being obscure; thus the face masks with the Dragon Gowns. The term "visible invisible" neatly misses the reality, for it completely conceals the HOSTING aspect. Indeed the HOSTING aspect can be particularly frightening, but is generally expressed simply with the baptism with water as H20; and using the numeral "0" instead of "O" to make the point of the Galactic connection.

So an advanced Galaxy is able to place into small child such of its designs as it wishes to check on the suitability of a location to visit.

No electrical technology, no digital technology, and it is unlikely that they would bother to send one of their officials. That certainly was the position RTTUR THNGME took when the request was made in 1952 to revisit the ATLANTIZ restrictions.

It is not a full three score and ten from that date, and while I am presented on television with many supposed "alien" visitors, I cannot be satisfied that I am not dealing with doctored images, especially those allegedly released by state media.

There is no sign of a Boat of Millions of Years, which would be a clear indication that the Double headed Serpent Mechanism is active.

I note a helicopter hit the Mull of Kintyre, and I am aware that checking of that " Double headed serpent mechanism" can create problems for local aircraft, but as with a causeway, the mechanism is only active at certain times and in certain circumstances. Currently the LOCK GATES in a Galactic sense are closed; the "Giant's Causeway" is covered with water. Besides that Gate is to THRENZ and the Lady KYEN TY AMEN TY has refused Activation.

≈ 19 ≈

Textual Revision.

At the time of the Reformation various local practises were checked by the RTTUR. Yes there had been some improvements, but they were patchy and pace of improvement was glacial. The stones were not exactly talking, and the locals were moving in a different direction.

RTTUR FVRRAKOTHERA was not impressed with the diseases which had infected his territories, and the clean up, had not turned out as such. Yes he and other RTTUR in that area had managed to hide most of the technology, and he put a BLOCK on accurate translation of the Maya script. As a block a re write of local texts was done with the *Popol Vuh*, so that again much Galactic detail was covered with incorrect dogma.

Various other areas had likewise had the population locked down with the loss of texts and understanding, particularly with the ANANN 00 KY, AH00RA MAZDA and NYMRYD, with AH00RA MAZDA marking out the Number 1 Galaxy who had provided the NYKYDYNYZZ material and that ANANN 00KY was the same Galaxy and that it was THNGME.

He had been a busy boy RTTUR THNGME, and he had also managed to put misunderstanding in place as an extra barrier. Something that the locals might have been inclined to pass off as LL 0 KY!

Again once the Galactic understanding is appreciated, the "0" Galaxy connection is relatively easily spotted; the difficulty is with the understanding that so many Galaxies are involved and the distances are enormous; unless you have some very specialised technology, like "The Boat of Millions of Years", which as MAWAHASATAS DA KAS can also be referred to as "My Chariot of Fire."

So in certain respects the Greek Pantheon of gods was a starting point, but it also had the capacity to conceal the essential Galactic Science, THAT the TWENTY FIVE RTTUR had designed all their Galaxies and set them in order, as a playground! So nothing which is hurtful or harmful is allowed on the playground of these very advanced PEARL HARBOUR. Some can have boundary walls so as to designate the garden from the jungle, and in that regard Hadrian's Wall, as it now stands, damaged, still remains as an unseen and unappreciated Galactic boundary marker and the 1707 Treaty of Union a Galactic SALT Treaty, in the modern understanding of a SALT treaty.

The Galactic RTTUR only restrict themselves with their selective amnesia, and it has to be remembered that Galactic Dragon; the RTTUR are all seeing.

≈ 20 ≈

A mighty Messenger with a little book

I might say the "AN DRUAITH" book of medicine, which is a little book, a bit like the Hippocratic Oath, in its terms, where the only medicine is to do no harm. Thus evaporated water allowed to cool and to fall as rain. And so the lakes and lochs are replenished.

Why would an RTTUR wish to take on a frail Atlantyon, Tyger, Wolf or Rhino frame? In the main to enjoy themselves on the FOOTSTOOL, and to "play with their friends," except that not all on the FOOTSTOOL are their friends. Lessons there for the RTTUR; which they well know!

On the advanced FOOTSTOOLZ, much building to be done, actual physical work, design, and after designing and building resting. Once every seventh day! Not a bad way to work things. No slavery, no beating, no mistreatment, no dominion.

In many respect a very different style.

II

"Do the locals want to know?"

The question is put by THNGME to his TWENTY FOUR ELDER RTTUR.

It is all one to work with the fiction, as with science fiction. It is entirely a different matter to work with the reality. And it is the Galactic Reality which I have been trying to describe with these my Atlantiz Texts.

The Reality I will cover in the next five chapters as I consider the Reality, with the assessment of FIVE of the RTTUR.

My understanding of the RTTUR is that they can appear as an ordinary individual, can HOST an ordinary individual, so that they may be satisfied that the FOOTSTOOL is a safe location to visit and to ADMIT the FOOTSTOOL to their GALACTIC system, even with territorial restrictions usually marked off by WALLS, even where damaged.

Their HOSTED individual they would term TZUTRAN GYDRAN (male) or ZUTRAN GYDRAN (female).

The position on the Ground must be safe, in their terms and where it is considered that it is unsafe or will so become unsafe REMOVAL of their GYDRAN becomes an INSISTED matter with LOCK DOWN IMPOSED.

I am well aware the RTTUR do not do gender, but insist on "the old fashioned way." SEX.

Thus easiest for them to speak when in Tyger, Wolf, Rhino or Atlantyon Shape.

For THE PROOF, my preference is for a Real Boat of Millions of Years to put in an appearance rather than the fake Unidentified Flying Object; particularly in these days of Digital fakery.

As understood, by the RTTUR there are certain specific matters that have to be covered by these particular aliens.

Thus I am aware that FVREKKRAN insist on very specific head style, so that pointed ears are out but speckles (as with speckled Kaymen) are in.

The same applies to teeth. There are specific styles even as with the Rhino styles, who can grow in missing teeth.

They are not talking claws, but ten digits in two hands, eight fingers and two thumbs, with nails not claws, and that applies to all designs.

The Tyger and Wolf designs have fur and some of it styled very carefully. The Rhino and Atlantyon have thin hair. The dog designs can be smooth as with X0L0.

TY MA HY HAAT and LLYKURRY0N, METHINKZ!

And that about sums it up!

Unless the locals here can accommodate the real Aliens in the correct styles, then no advanced contact. If the real Aliens are shown in a bad light, as with some science fiction, there will be no contact, no communication.

Equally if the locals here make daft rules and insist on Dogma rather than real science there can be no contact.

I have provided some ATLANTYON Language, and that is essential to speak with these aliens, regardless of the fact that many of them who can make contact will be fluent in English, Spanish and Mandarin. Certain languages cannot be spoken and should not be uttered or displayed as these are Lock Down matters.

This text will be a slim volume, even with the Technical Specification. It is somewhat longer than the MA0RI myth and legends, but it does contain the ATLANTYON.

TY TY HAKKA TY TY

K0NTTYKY.

The New Understanding

≈ 21 ≈

Backwards Going Forward

What Plato wrote about Atlantis I consider to be fiction. I had thought in dealing with a circular City, he was dealing with an understanding which might relate to a Galaxy, but it would appear that it was simply a story of a fictional land; something which could be aspired to; except that the conclusion was that it had been destroyed by the Gods.

Greece was the source of the story and there appears to have been a couple of candidates for his Atlantis which would have placed it in the Mediterranean.

If that is the reality, then the culprit is the THERA volcano, now called Santorini, and its eruption in BC circa 1600 did indeed cause the collapse of the Minoan Civilisation in Crete, and by comparative accounts the Tsunami that it caused hit the shores of Egypt, but with less devastation that in Crete. But that was long before Plato's time and there is no direct tie in by Plato to that historical reality.

The biblical pillar of fire is likely to have been the fire from this volcano, and as some have pointed out, before hitting Egypt, the water could have been drawn back to create a passageway out of Egypt, over dry land before the Tsunami hit.

There is also the possibility, that beyond the Pillars of Hercules there was another super massive volcano which blew its top, and the remnants of that are now the Canaries, centred on Tenerife, and the outer islands of that group being the rim of a super massive volcano. Possibly in times past, but likely to have been the Millions of years of history; nowhere near to Plato's time.

The ATLANTIZ I have described in my various text is a Galaxy matter, where this Galaxy is called ATLANTIZ_DRAAZ in a group of Galaxies, which for ease of reference I have referred to as THE KHRYZTAAL GALAXIEZ.

I deliberately use a different spelling to highlight that I am talking in GALACTIC terms, and not Globally; nor Flat Earth.

Evidence for such a hypothesis is shaky, and I can really only point to the Pyramids in Egypt and the Americas and the travel Mechanism I refer to as "the double headed serpent", the superlight travel mechanism which permits Galactic travel in a reasonable time frame.

In current Science fiction it is referred to as "wormhole travel", but the serpent reference, as I have said can be seen from the Egyptian wall paintings, and in the Americas with the Pyramid, which at certain times shows two serpent designs appear on the edges of the pyramid. And I must not forget the individuals in the stone carvings emerging from the mouth of the serpent. In the case of the Americas the individual would appear to be KWTLZ KO ATLZ. I use a different spelling to reflect what I regard as the Atlantyon Language, and the interpretation which I put on that is KWTLZ (a Star Name from a star in M64 Galaxy), "Ko" the Atlantyon word for mouth and ATLZ, the short contraction for ATLANTIZ, the FOOTSTOOL, which is this Globe, with the Letter Z being used as an identifier as ("read in a Galactic Context."). In my earlier books I had used triple chevrons, before and after words to represent a triple Galaxies interest and understanding, but was informed, including by my publishers, that they did not like it; and so on any rewrite of my books 1 to 5 the triple chevrons can be removed, provided the letter "Z" is added to make the point that what is written is meant to relate to real Galaxies and is thus not fiction.

In fact it does not matter if all my texts are viewed as fiction, because the reality does not change, and what I am attempting to explain is of vast Galactic civilisations, extraordinarily advanced and largely peaceful who reach out to the friends who are of compatible design, Male and Female, and to specific bi pedal designs.

That has been explained in various parts of the so called "Old Texts" I have written, in this book and previous ones.

The New Text deals with the reality of those advanced civilisations, and particularly their RTTUR figures, Heads of Galaxy Groups, all TWENTY FIVE of them.

While it may sound like the old Multi God System, there is one of the Twenty Five to whom the others defer, THE I AM THE; as I might refer to him. He is very far from Gender Neutral, but I am dealing with civilisations whose individuals can shape shift, into distinct and restricted styles, of ATLANTYON, TYGER, WOLF and RHINO, and in a tri colour style of individuals.

The Atlantyon can come in Pink, Dark Blue and Yellow, but that is merely skin colour. They can also come in Blue Skin, but that would be from a different Galaxy. Dark Blue, the locals here might call black and light colour brown for some. White might be used, but being technically picky, white is actually pink skin.(translucent skin, with red blood in the system design).

The other difficulty with this ATLANTIZ, is that it is all to do with design, from the globe to the species, and as a water based FOOTSTOOL, the reference for it is PP(PYKAAZ_ PYNWHAAL). I could say the sign of the Whale and the fish.

So the ATLANTIZ I am dealing with is enormous and with its canvas of Stars in this Galaxy, a "Golden Fleece" or a "Burning Bush."

I had forgotten the Nordyk and the so called "World Tree" and the star charts which accompany my texts could be said to look like a tree. In the NORDYK they refer to YYGDRAAZAL (albeit with a different spelling but similar sound, as if misheard).

My Old TEXTS cover that.

This text is about THE REAL ATLANTIZ , REVISITED, and the Hypothesis for that is that this Globe is one of their Galaxy FOOTSTOOLZ in this Galaxy, is visited from time to time by one of their officials, a KYDRAN, KURALAKYZ or GYDRAN to check if the population is sufficiently enlightened to be restored to this Group; and I have given weighting with silver coins, with various Sterling Coins representing the level of development and enlightenment.

Their assessments are done on a "Thief in the Night", "play with my toys" and "up close and very personal" basis, sometimes exceedingly difficult to spot.

They are very touchy feely, and the assessments delicate, thus in part my use of the name THE KHRYZTAAL GALAXIEZ.

So with this text, Book Seven the assessment draws to a close.

It was the local LANAZT Star who asked for the assessment and at 3d Ag 500, the weighting is below the level for contact.

INDEED matters are so bad, that there is a real risk the 3d weighting will be lost. That would mean the star would return to its LUU SY FHAAR (lost enlightenment) rather than the System Galaxy names as MR DRAGON ZYR and HUMMINGBIRD (male and female).

The assessments are very touchy feely and can depend on very minor matters, or not so minor matters, and may sometimes be referred to as ACTIVATORS to LOCK DOWN or to OPEN this Globe to Galactic Travel and Communication.

Book Six, THE LOST ATLANTIZ, drew attention to the fact that ATLANTIZ had been in system, but had been thrown out, and for a long period of time. It is about to be thrown out for 5,101 years, for continuing bad behaviour. It is far too violent, murderous, lacking in correct medicine and with unsafe legal systems.

INDEED, RTTUR MNZYVZ has pointed out that the test done at the time of the Crucifixion of KHRYZT has still not been satisfactorily resolved, and no apology has been given. MNZYVZ RTTUR rightly says why should his Galaxy continue to supply part of the NYKYDYNYZZ material(wrongly locally called dna), and draws to attention that MN0AN is actually a MNZYVZ term for the Cartwheel Galaxy connection, and that he can be very difficult.

He withdrew NYKYDYNYZZ material in WW1 and WW2 and is of a mind to instruct withdrawal; and of his technology and medicines. There is clamour for the other TWENTY THREE to do likewise, but with due deference the matter is left to the TWENTY FIFTH TT RTTUR, THNGME.

He raises the matter with MR DRAGON ZYR, because he is less than impressed.

"I was persuaded Mr DRAGON ZYR to appoint a KURALAKYZ GYDRAN on the basis that you have a territory which had a balanced and fair legal system, with no harsh and oppressive penalties and where the medical facilities would work the Hippocratic Oath, and where medical treatments were free at the point of delivery, and where the local Government collected funds for such, and where the taxation was not vindictive.

The material point was that my KURALAKYZ GYDRAN was not to be threatened in any way.

You were given GYDRAN on the basis that the GYDRAN could overrule minor failings, but it has been brought to my attention that is no longer the case. Harsh and vindictive penalties are imposed using the digital technology, and there is mass intrusion by the State of private data, and that in such data there is material which conflicts with our PROTOKOL, R18 RESERVED IN FAMILY, INSISTED BUT NOT FORCED. RTTUR LLOMOTTOR, of "THE BLACK ICE GALAXY" had been asked to install the digital technology and to monitor its correct operation; and to withdraw it, if it was misused by State Governments.

The LOCK DOWN to be imposed will be for 5,101 years from 2019.

I am very disappointed that it has ended this way, but I regard my GYDRAN'S KOMPLAINT, valid and accurate, so I am bringing forward the LOCK DOWN to 31 March 2024 AD 107, and after that date the Globe will no longer be entitled to AT 0N protection, and MY MANDATED GYDRAN will be withdrawn to make the point. I may defer that withdrawal to the 2030 date, but I am grossly offended, MR DRAGON ZYR at what I find on the restricted Reserved Territory to the North of Hadrian's Wall.

3d Ag 500 is a too generous weighting."

≈ 22 ≈

Walking in the Light

The expression is DRAGONSPEAK, which means that it means exactly what it says, but in a different context.

The Old Egyptian NYTYR were often shown with a sun disc on their heads,the Pharaohs with a golden cobra.

KHRYZT brought a New Enlightenment, but the locals wanted to impose their Dogma insisting on a Blood Sacrifice. On that Pilate complied. Thus the Galactic Test put in place when the baby had been delivered, is brought to a very swift end, far short of the Three Score and Ten years complete for the Galactic Test to be run, when the Crucifixion of the Messenger occurs.

I use a different spelling for KHRYZT (Christ) to make the point that I am dealing with that visit from a Galactic perspective; rather than from a local one.

The various Galactic Tests had been run since delivery of the Christ Child and various points are recorded in Matthew's Gospels. The difficulty is that while a local interpretation can be given to the events, the Galactic detail is far harder to see.

Why the flight into Egypt? To keep the child safe and it is the Messengers (Galactic) who arrange for that to be done.

In Egypt the child may be introduced to the correct teaching of the Galaxies, which was only faintly grasped by the Egyptian Priests.

Amenophis IV thought that his population could grasp his understanding, but how much did he grasp the Galactic Understanding? For a substantial period of time he was erased from history and it was not until 1922 when the tomb of TuatAnkhAmen was discovered that he was restored to the King's list.

It will be observed that I have used a different spelling for TuatAnkhAmen. That is done to reflect the correct Galactic Position, with "Tuat" being the word for Heaven (The Galaxies). Ankh is of course the symbol for life and AMEN a word very familiar in the Christian Tradition. I might use the term the "Mandate of Heaven" for the name. I could say, "of Heaven, the Living Lord", in essence the Mandated individual.

That would suit those who worked to the "Divine Right of Kings", but Cromwell put an end to that; except that the "Mandate of Heaven" is not a matter for the local populations, but rather for those who sit in Heaven; THE GREAT COUNCIL of the TWENTY FIVE ELDER RTTUR.

Of course such is my view and my interpretation of Revelation 5:14.

What I can see in 2024 AD 107, are enormous ANTI-KHRYZT populations, virulent and nasty. I see the Churches cowed by the mob, much like Pontius Pilate and the Courts likewise cowed by their political masters who have in many respects and in many territories infected the courts with political partisanship.

RESERVE AND REDAKT

The use of these words means that while I might fully explain, for reasons of safety from the mob, I am unable to speak freely. Such a Society where there is no free speech does not meet the GALACTIC PROTOKOLZ for communication and travel and thus the LOCK DOWN, POWER KLOSE is put in place and at a GALACTIC Level.

RESERVE AND REDAKT

These words simply serve to confirm that the GALACTIC PROTOKOL is not met and that The Galactic super light travel mechanism, to which the local population do not currently have access cannot be provided to them ; and given the State of the Nations they are not likely to have access for over 5000 years from now at best.

Where the population is clearly ANTI-KHRYZT there can be no access to superlight travel. That is the GALACTIC position.

No free speech, means no enlightened population, no advanced technology, and more seriously no advanced medicine. Antibiotics come into the category of advanced medicine; and the Galactic licence for the use of such has expired. It remains in place for existing antibiotics, but as for the advanced antibiotics, already withdrawn.

A population insisting on dogma, attacking the Galactic Visitor are clearly not enlightened. The Test before Pontius Pilate was for Pilate and Rome; and they failed.

DRAGONSPEAK, very difficult to spot, is with the "P P", as PYKAAZ_PYNWHAAL. Galaxy RTTUR watched that trial, and the upshot of that failure was to mark REKALL LOCK DOWN, and so that was in place until the King James Bible was allowed a print in the local English Language.

That would allow the Galaxy RTTUR to check for the ENLIGHTENMENT. That test was finally concluded and implemented on 31 March 2024 AD 107 when the imposed LOCK DOWN was confirmed as the local population had deliberately introduced ANTI-KHRYZT teaching and had attacked the GALACTIC understanding INSISTED of MILKY WAY.

SLAP DOWN in GALACTIC TERMS INSTRUCTED, as the Volcano in Iceland bubbles away.

The trouble with this Galactic Science is to do with events; and mainly that means VOLCANOES and TSUNAMI.

RTTUR MNZYVZ had pointed out to the GREAT COUNCIL that the same wickedness as had been presented to Pilate with the insistence of Blood sacrifice, had again been put in place by 2024 AD 107 and that was marked AGREED by THE GREAT COUNCIL.

The warning on that had been sounded but ignored.

RESERVE AND REDAKT.

During what was EASTER WEEK, 2024 AD 107 and at various times THE GREAT COUNCIL took careful notice of EVENTS to satisfy themselves or otherwise. Their tests; their checking!

RTTUR THNGME was satisfied his GYDRAN was THREATENED and with that the Local LANAZT Star as MR DRAGON ZYR had contravened the Galactic Assurances given.

≈ 23 ≈

OUR DELIBERATIONS

Following on the refusal in 2019 of THE GREAT COUNCIL to re admit ATLANTIZ, the matter was marked for appeal to the FIVE PANEL RTTUR, NY MY GURYON. The panel varied, depending on the Star and Footstool seeking re admission within a Specific Galaxy. The panel here are THE LORD FVREKKRAN, THE LADY AAGHAADDWR, THE LORD MNZYVZ, THE LADY ZYRAPPHAMA and THE LORD THNGME.

THE LORD FVREKKRAN is in White Tyger style, at a room in his palace in FVHAALYA. It is mid afternoon, the day is still warm and he is standing by a large open window looking over part of his city district towards the ocean in the distance. In the room with him are two male Tyger Guardians who have provided him with a written text, which he holds in his left hand. The conversation starts

i

Said THE L0RD FVREKKRAN to THE L0RD GUARDIAN,

"It is an interesting report and assessment, Lord Guardian. I am aware that you correctly understand that we have access to your local internet, and that we are well able to pick up material, even that which is not yet on the Internet.

I note you take some considerable interest on safe air travel, and I am aware that you are less than impressed with some of the local manufacturers. The safety of the crew and passengers is paramount.

The sticking point for me as RTTUR is what the locals wish of our Galactic civilisations. I am aware you wish the medicines from THRENZ, particularly the advanced antibiotics, and of course the matter transfer equipment to enable the blind to see, the deaf to hear and the lame to walk. In short you wish the real miracles on an industrial scale!

With our technology that does not present a problem, but the real problem is whether access to it is to be allowed.

I observe that in the Capital city, the streets run with blood. That is not acceptable. I have considered with LLOMOTTOR the misuse of digital technology to impose harsh and oppressive penalties. Given such I see no reason why we should supply more digital technology. INDEED I am agreed with LLOMOTTOR that much digital technology should be taken from you.

I do not need to go into the detail. You already know the detail.

On technical matters safety is important, and I cannot be satisfied that the locals would welcome the appearance of any of our Boats of Millions of Years. You are aware of our PROTOKOLZ ; that we do not wish to upset those civilisations which have not reached our level of development; and it is for that reason that we use and permit the display of Boats of Millions of Years; designed to look local and not threatening.

A Pearl Harbour FOOTSTOOL requires to be substantially peaceful, and that is not currently the case, as of March 2024 AD 107. There are two areas of major conflict. These as you know enable us to recall our material, including those designs which we have supplied, including such as to THYAM HYGLLANZ TY F0RTY.

FVRGHAAL has supplied material which meets our specification, and at a higher level than previous designs.

I am not persuaded, Lord Guardian to mark KONTENT to the re Admission of ATLANTIZ. I accept that the detail has been much updated but where I and the other RTTUR have to RESERVE and REDAKT to meet local sensitivities, there is no free speech, and that is not acceptable. I can view such lack of free speech to mean that I am looking at a Police State, with very dubious independent judiciary.

You appreciate that it is the RTTUR assessment which is vital, not what the locals say or think.

I AM required also to consider the routes the scientists and engineers take. You are well aware Lord Guardian that over the course of Millions of Years we as RTTUR are well aware of what works and what does not to provide the styles we require, both male and female.

As with THRENZ, and others like LLYKOTT LLYDRAN we do play a form of football and rugby, but style female does not lend itself to such as rugby or other contact sports for the simple reason that the females carry the young. This is a material design matter of very considerable importance, such that only with such as the matter transfer equipment can I change from Male Tyger to Female Tyger, to Atlantyon, Wolf or Rhino of the bi pedal kind.

I am afraid that with the Global weighting 3d Ag 500, not even 3d Ag 925+ I cannot begin to consider a Galactic Connection.

Those in State custody must be safe, and suicides and murders in prison means that ACTIVATORS are involved and we can deal with these SLAP DOWN. Such as the Chernobyl incident and Fukushima are correctly understood by you. The volcano ARRAN, slumbers, but does not sleep.

I am aware that you make difficult those passengers you require to find on a Boat of Millions of Years to satisfy you on the accuracy aspects, and you have my agreement that we will provide such; it can be done. Whether we are willing to do at this time, an dieser Zeit, that is another matter.

I am of the view that with the many deaths in State custody, including the so called Legal Murders, that I am required to mark ACTIVATOR AUTHORISON, and I so mark.

The technical marking from me is REKALL and NOT KONTENT."

FVRRAKOTHERA RTTUR

AMEN AM RTTUR

ii

THE LADY AAGHAADDWR is the RTTUR on the Panel, and her interest is for the female populations. She likewise is in her palace in AAGHAADDWR Galaxy, on an advanced FOOTSTOOL; also taking Red Tyger shape, which is a DRAAZ design. She is attended by two White Tyger female Guardians.

"Good Morning Lord Guardian. Perhaps I should say also good evening. My AAGHAADDWR Galaxy, is currently locally called "Silverado" and as you are aware I, as RTTUR am female.

You and I are both aware that we will not allow in the Satanic females, of which on the ATLANTIZ footstool, there are far too many at present. I noted you had some difficulty in clarifying the Satanic individuals, but you are correct it is a female designation.

You are aware that the style RTTUR AT FORMAL and for KURALAKYZ GYDRAN is with a face mask, or as you appreciate, with sunglasses doubling as a part mask. That is to correctly show the VISIBLE INVISIBLE understanding. The Dragon Gowns and FLYFL0TT also show that the VISIBLE INVISIBLE concept is correctly understood.

You wish me in WHITE TYGER mode, but given that we all do shape shifting, perhaps the Black Dog shape, as with ANUBYZ would be more appropriate. After all currently what I find at ATLANTIZ a very much Anti-White agenda. If I am being picky, white is in fact pink in skin tone, as the skin is very much translucent in the ATLANTYON design.

In this instance I choose RED TYGER as with SEKHMET. I am aware that is an ATLANTIZ_DRAAZ design, but given shape shifting it is one I can adopt, and one which I consider most appropriate.

I am most reluctantly agreed with the 3d Ag 500 weighting, and as you will appreciate I am in charge of many female aspects.

We all work to "Milky Way Practices", but there is considerable variation with such, which does not present as a problem provided it is done correctly, and in our case that means more than one male per female, but in strict order, Number,1, 2 and 3. That is primarily to cover a situation where Pilots of Boats of Millions of Years are away from our FOOTSTOOLZ for many years, and some individuals such as yourself, are away on our business for 70 plus years, as the FOOTSTOOL checking is run; and some do not return but move to other FOOTSTOOLZ.

Looking at events 2024 style I like FVRRAKOTHERA am not persuaded that contact is appropriate. I am KONTENT that there is a NOT PROVEN verdict but I have to mark as NOT KONTENT when there are show trials, where the weight of evidence is deliberately turned against males by the females; which makes the females to the style of Lady Macbeth, Salome and the Queen of Sheba; and are assessed by us as Satanic. The PROTOKOL is R 18 RESERVED in FAMILY, insisted but not forced, so State intervention under 18 is not permitted.

Like the Lord FVREKKRAN I take deaths in State Custody as an ACTIVATOR matter; and that includes legal murder and suicides; with particular reference to murder dressed up as suicides, or deliberate failure to provide medical treatment for those in custody.

It is we as RTTUR, including specifically THE GREAT COUNCIL at KAZTDATA meeting who determine who may be allowed to hold our MANDATE; and murderers are NOT so entitled.

I am of the view that with the many deaths in State custody, I am required to mark ACTIVATOR AUTHORISON, and I so mark.

The technical marking from me is WITHDRAWAL INSISTED REKALL and NOT KONTENT.I will give the ACTIVATION date as 31 March 2024 AD 107, and I have KONKUR on that date from THE GREAT COUNCIL."

RTTUR AAGHAADDWR

AMEN AM RTTUR

iii.

THE L0RD MNZYVZ

He is in Black Tyger design as he gives his assessment.

"I am not trying to be difficult, Lord Guardian, but as they say here in THRENZ, where I am visiting with the Lady KYEN at the hillside summer villa, "The truth is obvious, it is so obvious that it is seen only with difficulty, but is seen all the time."

The basic position for entry to our KHRYZTAAL GALAXIEZ is that nothing which is hurtful or harmful is admitted, and in that regard one of the criteria is a balanced and fair legal system which does not have harsh and oppressive penalties. I have deliberately left out the word "vindictive" because I am aware that vindictive is far worse, and for such vindictive systems penalties are applied.

You are aware of my position in respect of WW1, and even now in 2024 AD 107, the "ACTIVATORZ" are mostly hidden, but some events disclose ACTIVATION.

I had agreed to my design information being inserted into the males, but once it was manifest that they would be legally murdered because of my style, I naturally requested the return of millions.

Even now as you ask of THE GREAT COUNCIL for the release of vaccines and advanced technology, because of Legal systems which are not balanced and fair, and in certain instances are clearly vindictive, we will not assist.

It is not sufficient to say, "I find no fault" and then hand over to be murdered an individual who is to my design to be murdered judicially in order to uphold "the rule of law" which requires murder.

I skip around the matter, but like the LADY AAGHAADDWR, I do take note of "deaths in custody" and manifest bad law. As the saying goes, if they live by the sword, they die by the sword. You Lord Guardian have referred to that as the mirror rule, and to a certain extent it is.

Like the Lady RTTUR AAGHAADDWR and the others of this NY MY GURYON Appeals Panel, I have to consider ACTIVATORZ when I am dealing with the so called "Divine Wind", the 1335 days of Daniel 12:12.

The main purpose of the ACTIVATORZ and the 1335 plus days, is to delay the delivery of those weapons, which some refer to as "mushroom clouds" but which naturally I refer to, as do you, as "DRAG0NZ TEETH. It is very convenient for the locals to manufacture and store such, but as you are well aware accidents can and do happen. Mostly we try to avoid accidents, but when the legal system is vindictive, especially where it murders those in custody, whom it should protect, there is the opportunity for an ACTIVATOR to become operative. The brakes, so to speak, are suddenly released.

I am aware that both RTTUR FVREKKRAN and RTTUR AAGHAADDWR have marked NOT KONTENT to the re admission of ATLANTIZ as our PROTOKOLZ have not been met. It is actually far worse, and the Local LANAZT STAR, as MR DRAGON ZYR has permitted on our reserved territory the worship of the MURDER KULT. I refer specifically to the TUNDERGARTH incident.

I am aware of the KOVER UP with such; which was part of our Judicial Checking Exercise. Den Haag as you are well aware is, like Carlisle, not in our RESERVED territory.

I am going to do a RESERVE and REDAKT as the specific points have been covered by RTTUR FVREKKRAN and RTTUR AAGHAADDWR.

You observed that I have adopted Black Tyger shape, as it would appear that on ATLANTIZ, in our reserved territory it is local practice to offend against white. That is unacceptable.

I am going to take the date of 31 March 2024 AD 107 to INSIST on a NOT KONTENT and to refuse ATLANTIZ re admission. The LOCK DOWN will run 11:150 from midnight 31 December 2030 AD 107; but we will have marked our AT 0N WITHDRAWAL on 31 March 2024 AD 107. Midnight!

My ACTIVATOR AUTHORISON is with the lack of Free Speech in our RESEREVED TERRITORIES and I so mark IMMEDIATE."

RTTUR MNZYVZ
AMEN AM RTTUR

iv.

THE LADY RTTUR ZYRAPPHAMA is in her principal city, in what is locally called the Sunflower Galaxy.

"As you would expect, Lord Guardian, I am one of the PANEL NY MY GURYON.

I have examined the reports carefully. I am agreed with the others of this NY MY GYRYON panel, who are considering the appeal against our refusal to re admit ATLANTIZ.

Like myself, the Lady AAGHAADDWR has drawn to our attention that various States adopt a Murder/Suicide policy, to those in their care which put them outside our protection.

Suicides in respect of under 18's is simply, not acceptable, in the territory to the North of Hadrian's Wall. That territorial marker is of course Galactic, not Global, having regard to our Galactic position.

I am aware also of the reintroduction of Roman rule, or English Rule as it is sometimes described; but both of these relate to genocidal philosophies, most recently seen in the period from the 1930's. I am aware of the KAZTDATA position in 1952 AD 107, and furious anger of RTTUR THNGME at the desire of the locals to persist with the Murder Kult to the genocidal extent.

My ACTIVATORZ as you well know relate to the Murder by the Government of their own people; and in one particular instance of the murder/suicide instructed of naval cadets. That was the LAST ACTIVATOR, which made the matter IMMEDIATE.

IMMEDIATE in RTTUR terms can involve an element of delay; in that instance of only a few days.

I also dislike rendition policies, particularly where these relate to long custodial sentences likely to lead to suicides/ murder.

Mr DRAGON ZYR has a large number of teeth which can make the point. We THE TWENTY FIVE have to authorise, and I note that RTTUR MNZYVZ has moved from AVIZANDUM to IMMEDIATE INSISTED.

I have reached my decision, and I can take it on an extradition matter and on the NOT PROVEN matter. That verdict NOT PROVEN serves as useful purpose to protect from malicious prosecutions and to demonstrate that GRACE is available. I will use the cancellation of that verdict as the ACTIVATOR, reserving to RTTUR THNGME the matter of the EQUATOR VOLCANO, which he has already marked as INSISTED. That ties neatly with RTTUR AAGHAADDWR

So my ACTIVATOR I will take as RENDITION/ EXTRADITION to countries who have failed Legal systems; of those who worship genocide.

If they wish Apocalypse it is in the local hands, but I mark NOT KONTENT to the re admission of ATLANTIZ. I do however insist on the removal of THE LORD GUARDIAN and the return to THRENZ or to THNGME; our Galactic doors are firmly slammed shut, LOCKED BARRED and BOLTED.

Malicious prosecutions, as you well know can lead to slap down and you are well aware of that. The ACTIVATORZ means that the BLITZKEIG can fall where it last fell, although my preference is one large CITY and one HOLY City; most likely as a result of an accident! Accident includes the misfiring of a ballistic missile because of a communication mix up and the failure of both a self destruct button and an intercept failure.

You are correct Lord Guardian that I instructed the recall of Millions following on the Titanic incident; a slave ship by design, and the incident with Hindenberg at Lakenheath, making an unseen locally similar point.

Safe civilian air travel is insisted, and we do not like cover-ups on point. There is insistence on full disclosure and by 2030 at the latest. I am not holding my breath on such, but in the absence of such disclosure on the matter of the re admission of ATLANTIZ I mark NOT KONTENT INSISTED.

AAGHAADDWR and myself are NOT KONTENT about the treatment of women on the FOOTSTOOL ATLANTIZ.

I likewise have read the report and I am going to do my assessments on the prison incidents which led to suicide. I snap the ROODS so that the connection with my Galaxies is broken. As before we will collapse a bridge to make the point! The Golden Gate may be said to be the easiest to snap, given that we have already done a display at Takoma. DRAAZ can do a snap at TAY as was done previously to make the point, but that is a matter for the Lord TAY HAY TY. Personally Since I am making the move, the snapped bridge has to be in the Americas.

Given that I have marked ACTIVATOR AUTHORISED, I likewise am doing a NOT KONTENT, which I think leaves the matter with RTTUR THNGME.

The activation date is confirmed 31 March 2024 AD 107. FINAL and INSISTED; as with August 1945 AD 107."

RTTUR ZYRAPPHAMA

RTTUR AM AMEN

V

THE RTTUR THNGME

The matter was INDEED with THE LORD THNGME. He had appointed a GYDRAN to run the Assessment, THREE SCORE AND TEN complete, and that would be reached late December 2024 AD 107.

He was well aware of the KHRYZTAAL GALAXIEZ PROTOKOLZ, and it would be fair to say that considering the report and the appeal by MR DRAGON ZYR, RTTUR THNGME was not best pleased. Politely it might be said he was in KREAM PUFF, a specific DRAGONSPEAK term which indicated substantial displeasure.

The GYDRAN had assessed at 3d Ag, 500 which meant that ATLANTIZ was not re admitted, and the LANAZT Star had appealed that assessment, and thus this NY MY GURYON panel.

RTTUR THNGME had carefully considered the assessments by the other RTTUR on this panel.

Technically the 3d Ag 500 assessment meant that the LANAZT Star and its FOOTSTOOL were in System, but with no communication and travel links. Basically that meant in practice, no change.

The major local Governments might look for the Aliens but would be unable to find them. Not that they were not out there in the so called "Bottomless Pit", which was one way of describing the vast Universe.

The GYDRAN had explained the position locally, but on the local dogmatic view, as with Flat Earth; there were no civilisations out there. The LANAZT STAR was alone in the vast empty universe.

Of course the distances were vast, and rocket science would take the locals nowhere!

What was required was that the superlight travel mechanisms were made available.

There was no point in taking the matter to Avizandum as MNZYVZ had initially done, and then changed that to a NOT KONTENT, with ACTIVATION. There was a Galactic Position at

Play. He could mark KONTENT and over-rule the TWENTY FOUR, but why should he?

He was not Portia and this was not the Merchant of Venice. The "Father forgive them," did not work either, because as with that incident the locals were fully conversant that they were asking for and insisting on a blood sacrifice."

That brought him back to that TUNDERGARTH incident, and to the Twin Towers which followed. 11:9 was an interesting reference. Area 51 meant to him the FVREKKRAN Galaxy, and he had seen the view taken by FVRRAKOTHERA. Options were limited.

He could leave them with Plato and the Atlantis of myth. Technically the locals were walking in darkness having given up on the light, and that aspect had been confirmed by the prison alleged suicides.

He was very annoyed. His KHRYZTAAL GALAXIEZ he would continue to keep well hidden, and that meant NO KONTAKT.

He was aware their assessments were "Touchy feely" and all the TWENTY FIVE RTTUR were aware that when they appointed a KURALAKYZ especially if it was a GYDRAN, that it was for the KURALAKYZ or GYDRAN to report.

One critical aspect was that when appointed, the GYDRAN was not be threatened in any way, and had to be able to speak freely, without threat or hindrance. What the locals referred to as "Free Speech".

THNGME was aware the GYDRAN KURALAKYZ was making the point that RESERVE AND REDAKT had to be applied even when addressing THE GREAT COUNCIL.

The GALACTIC message had been left in a very stylised form, as THE GREAT COUNCIL required.

The 3d Ag point 500 level meant that the locals could not accept the Galactic reality, and for the most part were insisting on local Dogma; especially big bang theory and evolution; neither of which Dogmas were acceptable in Galactic Terms.

There was also the difficulty about the "before Moses" and the "I AM THAT I AM", which was the THNGME marker, and as RTTUR MNZYVZ had already pointed out, despite the 2000 year time difference, the locals insisted on their local dogma even when scientifically it was incorrect, with man-made global warming being one of the dogmatic blindness aspects.

RTTUR THNGME was aware that RTTUR LLOMOTTOR could deal with many on the FOOTSTOOL using his RED RHINO, but such would not be required.

The assessment was with THNGME. The purpose of the TWENTY FOUR ELDERS was to assist THNGME with his deliberations, but he had to have regard to the billions of their safe secured FOOTSTOOLZ within THE KHRYZTAAL GALAXIEZ. He had noted his other TWENTY FOUR RTTUR had confirmed the 3d Ag point 500.

The Globe would be left as it was, but the display of the DRAGON GOWN, with ATTYFF and DRAGONSTICKS remained Restricted.

In many respects as assessed by the other four RTTUR on this NY MY GURYON PANEL, there had been a number of serious PROTOKOL contraventions which simply could not be ignored, and put bluntly THAT meant that the assessment decision was taken at midnight on 31 March 2024 AD 107.

It was self evident that the locals indulged in a vindictive Justice System, with harsh and oppressive penalties, that not only was there no free speech, but the Legislation enacted had more than a chilling effect; it put conversation into the deep freeze and all to preserve Dogma over Galactic Science; PREZAKTLY the same test put to Pontius Pilate and which had been failed at that time and now failed again.

RECALL of the GYDRAN was instructed IMMEDIATE, subject to the discretion of the GYDRAN, the MANDATE OF

HEAVEN, which was in the Gift of the GREAT COUNCIL was removed from the FOOTSTOOL effective midnight 31 March 2024 AD 107, ROTATION of individuals would be further restricted to THUURRNN GALAXIEZ, with many now being directed to LLYKOTT LLYDRAN, and to LLOMOTTOR'S Red Rhino.

The decision had to be taken now. The last ACTIVATOR had been put in place. MR DRAGON ZYR was released as of 1 April 2024 AD 107 to deal with matters; including using the DRAGON TEETH to complete the tidy up to KHRYZTAAL GALAXIEZ specifications.

To the readmission of ATLANTIZ, at this time THAT was REFUSED INSITED, and for the matters stated.

THE RTTUR THNGME MARKED NOT KONTENT INSISTED.

AMEN AM RTTUR

≈ 24 ≈

MR DRAGON ZYR REPLIES

RTTUR THNGME did not expect an immediate reply from MR DRAGON ZYR. The point had been made in usual RTTUR STYLE, somewhat surreal, like a soft watch marking the LOCK DOWN as INSISTED; and with the SUNFLOWER connection once again snapped. It was not what he had wanted, but MR DRAGON ZYR had been warned from the outset in 1952 AD 107, that the GYDRAN was not to be threatened. Yes the GYDRAN could take an assessment, and could waive the matter away, and had done so on several occasions. It was NOT for MR DRAGON ZYR or HUMMINGBIRD to Test the GYDRAN, because if they did such they were technically threatening at GALAXY level and such was a DRAGON'Z TEETH matter.

ii

Said MR DRAGON ZYR to the NY MY GURYON panel,

"It is accepted that your KURALAKYZ GYDRAN is threatened, but in appointing a GYDRAN in the first place you expected that there would be challenges and threats, which The Lord Guardian could waive away.

I do accept that calls for murder are compltely unacceptable, and you are well aware LORDS RTTUR that I am well able to deal with such on a KHAARINGTON Style, which is not PREZAKTLY obvious to the local population; PP TAK AZKO. As you would agree Star matters.

Your Guardian is in place, and is not currently much disturbed although certainly irritated. The THREE SCORE and TEN is not currently completed, so that there remains time for rectification. That will not be the case if there is an INSISTED IMMEDIATE REKALL, but given that has not already taken place, even if instructed, perhaps a little more time. After all I am sure that RTTUR LL0KY, might enjoy a little amusement; a little bit of malarkey."

At that point MR DRAGON ZYR was interrupted by RTTUR LLOMOTTOR.

"Have a care MR DRAGON ZYR. This matter of RE ADMITTANCE is a GREAT COUNCIL matter and we are exceedingly touchy. It is the GREAT COUNCIL position that Pontius Pilate was given a "GET OUT" but chose not to use it. You will appreciate my bad grammar, but then I do bat for a Grammar School. VIA, VERITAS, VITA is insisted, as you fully appreciated. RTTUR LLYMOGREN can clarify if needs must, but I do not expect that to be necessary."

The reply was relatively swift from MR DRAGON ZYR.

"I appreciate and accept the SNAPPED CONNECTION, which is with THE LADY RTTUR ZYRAPPHAMA, M63 Galaxy, and I am sure in the particular circumstances THAT also means that YOU LORD FVRRAKOTHERA of FVREKKRAN, M51 Galaxy as locally called, and THAT RTTUR LLOMOTTOR of LLYKOTT_LLYDRAN, M64 also makes the point.

At issue here is not so much a Galactic Bridge, but rather a Galactic Tunnel. If I am to be re established with THE KHRYZTAAL GALAXIEZ, the so called wormhole connection is required as a reality. Yes my local population can do the fiction, but the reality requires your move"

The Lady AAGHAADDWR intervened, to interrupt MR DRAGON ZYR.

"Not so fast, MR DRAGON ZYR, nor quite so glib!

You have had WW1 and WW2, and while we have no desire for WW3, the point is made by our RTTUR THNGME, that the poison ought not to have been introduced to our and his territories, north of Hadrian's Wall and that on our Galactic Understanding, Rhine-Donau, South Shore Black Sea, South Shore Kaspian, the Wall at the Middle Kingdom that was, North Nyppon, Kanada, (Some Men Hate Eating Onions, south shore) and across the ATLANTIC, down to the KANA RYZZ and back to the GIANT'S CAUSEWAY, which is the marker for accessing the Double Headed Serpent mechanism; PP AD YZ identifier. We are NOT agreed to the display of a Boat of Millions of years, an dieser Zeit."

The interruption had been very technical, and the use of German made the star specific point. The reply was immediate.

"I AM obliged, LADY RTTUR, and the point is well made. Where is the LADY URZA on this?"

The reply now was from the LADY AAGHAADDWR, who was clearly now displeased.

"In RED TYGER shape, I speak also for the LADY URZA AM UR.

Our GALACTIC mandate cannot be held by those who worship and instruct murder, and such are refused from our SYSTEM GALAXIEZ, with the possible exception of restricted parts of the LL system. They may be allowed to see the "Promised Land" from afar, but will not be admitted.

While it is accepted that a tidy up has to be done in our Reserved Territories, Judicial Murder is NOT acceptable; and free speech is INSISTED. Not always obvious, but it is insisted. FVULL GAK0V! B and V, V and A!!!

This is a very old point MR DRAGON ZYR which MUST be respected. Your assessment, very generously given by GYDRAN THNGME is 3d Ag 500, not 1/-Ag 925 plus, and the 1/- Ag 925 + you are only allowed as GYDRAN THNGME is VISITING, and his safety is paramount, and as you accepted, he is threatened, that must be RECTIFIED IMMEDIATE.

Lies and deceit, especially on the part of Local Governments constitute a threat; a KLAAR and PRESENT danger."

She paused, and there was virtually an immediate response from MR DRAGON ZYR.

"I am obliged, My Lady RTTUR. I am very much in the hands of THE GREAT COUNCIL.

Some of your INSISTED TIDY UPZ are difficult and unsettling, even to me. I am in communication with the GYDRAN, and as far as I am aware he will UP WITH THIS PUT for short time on the basis that I have the matter in hand for a satisfactory resolution, IMMEDIATELY, in DRAGONSPEAK terms."

THAT brought the conversation with THE GREAT COUNCIL APPEAL PANEL to an end, an dieser Zeit.

27 March 2024 AD 107!

≈ 25 ≈

SNAPPED CONNECTIONS

Soft watches and Time

Galactic connections are often marked on the ground by Bridges and Tunnels. They are not obviously marked as a GALACTIC ACTIVATOR but the RTTUR well know where each one is located, and when new or replacement ones are put in place.

The TAY rail bridge was one such ACTIVATOR, linking the connection to DRAAZ. When the bridge collapsed, the connection was not re-established with the repair, but only when the Road Bridge was put in place. It is all a GALACTIC matter and UNSEEN unless clarified as such.

So that effective, 31 March 2024 AD 107, a Galactic SLAP DOWN has been put in place and MARKED OFF.

ii

The matter was with THNGME GYDRAN, who was fully aware THE GREAT COUNCIL were monitoring very closely. He did not require the Pontius Pilate story repeated *ad nauseam*, but he had acquired the BULGAKOV text as Pilate had been mentioned. He did wonder what THE LADY URZA made of it.

"Interesting" he noted was the reply from THE LADY URZA, who continued, " and the various strange names confuse you as with B and V. The Cyrillic font you find difficult to read. It is all Greek to you."

He could see THE LADY URZA smiling showing part large incisor, not as he might have expected in White Tyger shape, but in the golden AM UR. Surprising how the obvious concealed so much.

The matter in hand, and it had been with him for some time was the REALITY of THE KHRYZTAAL GALAXIEZ, and allowing for their PROTOKOLZ, whether the connection could be RE ESTABLISHED.

Holy Week was the appropriate time to consider that, and on that he had to agree, unfortunately, with RTTUR MNZYVZ, that the progress over two thousand years had been glacial. INDEED Glaciers might be said to move faster.

The RESURRECTION was the GALACTIC matter. So easily done with a Boat of Millions of Years, but hidden from the local population. And Thomas, doubting, how difficult were the Galaxies with that? All very specific! The Ethiopian eunuch likewise!

WW3 was to be avoided, but the GYDRAN was well aware there were numerous hidden ACTIVATORZ. In certain respects it was just when the last one was tripped.

Some he had spotted. Rendition of individuals, especially on an Extradition matter was exceptionally dangerous. RTTUR FVRGHAAL was watching on THAT one! Not PREZAKTLY obvious, but if carefully considered, exceptionally obvious, especially as a warning had already been given.

The difficulty was that when THE KHRYZTAAL GALAXIEZ could not be understood, the clock was at 5 minutes after midnight. It was INDEED a soft watch, probably in the hands of a white rabbit carrying a 10/6d ticket! The locals might be looking through a glass darkly, seeing the Galactic Plates, and missing entirely the UP KLOSE AND VERY PERSONAL. Amenophis IV and Tuat Ankh Amen!

The white Tyger RTTUR THNGME was enjoying his stay at the mountain retreat of THE LADY KYEN at South THRENZ, relaxing, but continuing to keep a very close eye on that ATLANTIZ matter.

He was in a very playful mood; and up close and very personal.

Technical Specification

I cannot be certain the locals here wish an updated Galactic Understanding; particularly where it conflicts with the various local Dogmas. On that basis there is nothing more to be said.

The Galactic position does not change especially when local perceptions cannot accept that. They are stuck with "Flat Earth."

This part of the text is supposed to represent the accurate and enlightened understanding. Essentially the correct update understanding of the reality which is this Globe and the local Star.

With the previous six books I have attempted to upgrade, conscious that with various earlier books I had made errors in the understanding which required rectification.

The last number SIX was called "The Lost Atlantiz", with the spelling changed to reflect correct Galactic understanding.

It is of course my view and assessment. It is not forced on anyone else because that is part of the correct Galactic Understanding. A critical point!

So the Hypotheses here, is that the Real ATLANTIZ, is the description of a Footstool in this Galactic Disc, which is known as ATLANTIZ_DRAAZ; moniker "The Golden Fleece" at one level, and at a very advanced level as "The Great Bluu Whaal."

Different spelling is used to distinguish the similar pronunciation from the Blue Whale mammal.

And again the understanding is that this Footstool is in the PYKAAZ_PYNWHAAL style, of the whale and the fish. PP for short; and at an advanced level of Understanding some Galaxies like those letters to be in the Galaxy Name; all explained by me in previous books!

As to the PROOF of this understanding, although I would contend it is correct,THAT is NOT PROVEN; and in many respects the people of this time, while looking for Unidentified Flying Objects, do not wish to have it proven that there are vast Galactic Civilisations in space, many of them thousands if not millions of years in advance of this one.

The calling cards of these very advanced Civilisations, I would contend have been left, in various locations as various myths and legends making reference to Heaven, but only in an abstract way.

Given that Plato wrote the story of Atlantis, it is my assessment **that Plato's Atlantis, was all fiction.** He did not know what lay **beyond the "Pillars of Hercules", or the width of the Atlantic** Ocean. As for the Egyptians, they knew as their Papyrus boats had crossed it, and had made their way back by way of Norge. The proof is in their Mythology, and of course the YYGDRAAZAL reference; even if I spell it differently.

My interest is in the DRAGONSPEAK; the Galactic detail, most often left in a hugely stylised form **and Plato's Text does not** conform to the Galactic Science style of, Preface, Old Text, New Text and Technical Specification.

The Angels in "shining robes" referred to in Matthew's Gospel at the end has a specific DRAGONSPEAK meaning. The Spartans knew what that meant, and in the Biblical text there is reference to the "Armour of God". I could say here put on the DRAGON armour.

The gardener was not immediately recognised for who he was and is, and while the predominately white and pale blue dragon gown can be made of a material like rayon, which will have a shiny appearance, particularly if illuminated by a powerful light; as with sunlight, or a very powerful light beam, the reference to a gardener would not suggest such DRAGON GOWN was worn, rather some other style of clothing.

As for the transfiguration that rather looks like the operation of the matter transfer and transiting equipment which is very advanced equipment, but equipment which would be available to "The Boat of Millions of Years." Technically it is a spacecraft which uses wormhole technology, and as it name implies can land on the water, and submerge, and could easily be described as a large white whale in the midst of a storm!

The proof of the Galactic Interest is all neatly hidden on the Egyptian wall paintings, including in particular that of "double headed serpent" with the NYTYR in his boat being pulled by six slaves. That would be pilots who are not slaves, and as the stylised Egyptian drawings are in profile; that would be twelve pilots. The wall paintings are an attempt to understand these very powerful Aliens; and I am satisfied with the NY TY R term for these "messengers."

Are my books nothing more than guesswork? As I said I have not been able to prove anything, because in this time frame the real aliens have not been seen to have visited. If any Government has the evidence, they are not sharing it with the public.

Another very difficult aspect to comprehend, but which is essential in this Galactic Science, is that the Local Star is living, can and does chat with individuals on this FOOTZTOOL, often in DRAGONSPEAK, and can also chat with THE GREAT COUNCIL, and can take action on this FOOTSTOOL setting off volcanoes, and Earthquakes, and very much more. The Star technical must thus be grasped by the Local population, and the local star is well able to stretch forth and protect those whom are to be protected; and that includes visiting Galactic Individuals doing the "Thief in the Night", "Play with my friends" and "up close and very personal" Galactic assessments for compatibility.

There is no point in leaving THE GALACTIC KURALAKYZ or GYDRAN in place for 70 years, when at age 35, the local mob have demanded a blood sacrifice, and the local authorities are too weak to resist bloodthirsty and genocide in the mob demands.

As for the air crash dummies produced as evidence of Aliens, they are only the equivalent of car crash dummies; lacking fundamental and clear correct detail.

Alien abduction is possible, but why bother especially when the group of Aliens I have described have matter transfer and as they designed the NYKYDYNYZZ material, loosely referred to a "dna", they can appear as entirely human; although with their advanced understanding they are referred to as ATLANTYON.

Across in the Americas, as with Egypt, there are Pyramids, and those in the Americas even have one which, at certain times of the year shows what appears to be two serpents; as with the double headed serpent mechanism; and for good measure shows individuals emerging from the mouth of the serpent.

As with an aircraft, a large civilian airliner, it is dangerous at take-off and landing, but which should be safe in flight. In Galactic Science terms it MUST be safe in flight, and at take-off and landing and in the case of the Boat of Millions of Years the mechanism compresses enormous electrical and gravitational forces to enable the craft to travel 15,000 light years in 5 days, from here at ATLANTIZ to DRAAZ and the 2.2 million light years in around 12 months from here at ATLANTIZ to THRENZ,(M31 Galaxy); and thus the safety aspects are paramount.

Certain very advanced civilisations can do the travel virtually instantaneously, but as they enjoy travel, there is no need to rush, and they may prefer the "Slow Boat." Safety is paramount.

So I have given in the previous books details of the Understanding which goes with this very advanced group of Galaxies as they select those who are compatible to be admitted to their KHRYZTAAL GALAXIEZ.

The group is mentioned obliquely in Revelation 5:14 as the Twenty Four Elders who defer to the Twenty Fifth, and whom I have taken to refer to as RTTUR; Heads of Galaxy Groups. In book six, I allotted them names, conscious that they could change the names to whatever they liked, save that one of them, the Head

RTTUR who referred to himself as, "I AM THAT I AM." He of Exodus 3:14; but he was not happy with Moses, and to a certain extent Moses was not happy with him.

I say him, but in the shape shifting context of the 2/6d Ag and above weighted civilisations, he can become she, but there are very strict PROTOKOLZ on that, and it is personal preference of the RTTUR as to male and female.

In a 3d Ag 500 civilisation, which this one is, the individuals cannot change sex, and if that is done using surgery or with drugs, then it is not only the Hippocratic Oath which is broken, but the Galactic PROTOKOLZ and KONNECTIONZ.

That is where I am with my text of THE LOST ATLANTIZ. The Local Star as HUMMINGBIRD and MR DRAGON ZYR had asked for the links to be restored. RTTUR THNGME flaked out at the suggestion when it was put forward at the KAZTDATA meeting in M31 THRENZ in 1952; book six reflects that position.

The review was authorised, and thus ATLANTIZ was revisited by a GYDRAN KURALAKYZ. Certain matters had to be advanced, including specifically safe civilian air travel, in comfort and at a reasonable cost. It was all going very well until 2019, when the assessment was completed with a NOT KONTENT. All TWENTY FIVE RTTUR so marked.

An appeal on that decision was requested, and the LOCK DOWN for 5,101 years was marked as 11:150, which meant a period of 11 years grace would be allowed,(assuming the Locals did not regress as with vindictive legislation) after which LOCK DOWN would be in place for 150 years, until next review.

ii

Fiction not fiction?

My ATLANTIZ texts are viewed as fiction. That is convenient.

The intention is that they disclose a reality, but as I indicated in the earlier Old Text, the New Enlightened Understanding is that it is difficult to send smoke signals to a super computer, unless of course I am talking about a high temperature supercomputer, a living individual, which is what the local star is; as MR DRAGON ZYR or HUMMINGBIRD.

Currently, in terms of this Galactic science this is a MR DRAGON ZYR phase; a bit like 1901 to 1952.

Amenophis IV of Egypt thought that he had grasped this Galactic Understanding, but in all respects MR DRAGON ZYR has to be satisfied, as does HUMMINGBIRD, even if it is not a HUMMINGBIRD period. Change of name to AKHENATEN,(rather than AKHENAT0N) does contain the HKT for the DRAAZ reference, but it is missing the AT 0N; and the omission is not simply a spelling error, because the "E" or "YY" if pronounced as "E" fails to record the "0"for the Galactic Disc and the Footstool which is contained in the 0N, where the numeral "0" represents the Galactic Disc. Milky Way as I have explained many times is a species design (and AD is abbreviation for ATLANTIZ_DRAAZ). By all means say "in the Year of our Lord" if referring to the individual who is RTTUR of this and many GALAXIEZ.

Son of God fine, but then that could be written in Galactic style as YYG ZYVZ; pronounced as Jesus.[The J being the ATLANTYON "Y2" or "YY"], usually pronounced as an emphasised "E" sound.

In terms of Matthew's Gospel, there is clear evidence of a Galactic Checking, with the arrival of the child and up to the resurrection; much to the annoyance of the locals who were unable to grasp the test.

That failure to understand the Galactic Test is where I am with this text. The position has not changed in over 2000 years.

The Galactic Internet

One matter which I had only faintly understood was that these very advanced civilisations needed the local Internet to be built here, primarily as a means of checking the development of the locals. Style wise, most of the locals are Human, and not ATLANTYON, and thus cannot travel on a Boat of Millions of Years.

Indeed, WW1 and WW2 was the removal by certain Galaxies, of the mainly male NYKYDYNYZZ material they had provided. It was being abused, so withdraw it on mass.

The mobile phones and electronic tablets were developed from Science fiction technology, making real the fiction of the mobile communicator.

As I carry around in my pocket a smart phone, I am aware that it can be used by a Boat of Millions of Years to make contact, and with my agreement such Boat can easily remove me from this Footstool.

As a "disappeared" the Galactic Science is only proven to me, the remaining billions proceed on with their lives completely unaware of these vast Hidden GALAXIEZ and their FOOTSTOOLZ.

The disclosure of these Galactic Civilisations is tricky. There is the desire not to interfere with the locals, and also an understanding that the locals here are not on the ATLANTYON wavelength, with their local philosophies and Religions. The Truth may be obvious, but billions prefer Dogma. When that Dogma becomes genocide, when the local population indulge in show trials, deliberately murder prisoners in their custody (including arranging accidents and arranged suicides) then these advanced Galaxies can instruct WITHDRAWAL LOCK DOWN; thus the new lock down for 5101 years to see what happens.

The local LANAZT star has obtained an 11:150 Review and currently it is with the Local Star MR DRAGON ZYR to upgrade or downgrade from the 3d Ag 500.

The request for advanced medicines and the wormhole technology is refused at this time.

iv

The Galactic Science

The reality does not change. It is obvious with the various Galaxies. The problem is that the local Dogma can hide the reality, and sometimes that is used as a LOCK DOWN matter.

With these now SEVEN books I have attempted to provide an explanation. I fully accept that it is my view, and I am well aware of the demand for proof. The KHRYZTAAL GALAXIEZ do not respond to demand; except to withdraw and maintain their barriers.

They have to be satisfied that the Globe is safe for them to PLACE THEIR FEET, and on their DESIGNATED TERRITORIES.

The design of this FOOTSTOOL is that many areas are technically unsafe, but the test is run on the areas designated as RESERVED. The assessment is very touchy feely and a minor infraction may result on the imposition of LOCK DOWN, POWER KLOSE.

The assessment has to be done by one of THEIR OFFICIALS; thus THE GYDRAN. How that is operated is a matter for them, but is well known to the local LANAZT Star.

From my point of view it is self evident these very powerful Aliens have visited in the past and will continue to do so.

They will only deal with enlightened populations, and the record of their visits is usually provided by the locals. They can be exceedingly difficult about the record left.

I have largely taken the record of their visits starting with the King James Biblical version, and its style. The Galactic point is that it is only with the publication of The King James Version, that these Galactic Civilisations agreed to THAT as a record as UPDATED. Thus on my current understanding the figure of Jesus as updated is

YYG ZYVZ. On that basis it is relatively easy to see how the Greeks could accept the new teaching; as coming from the "Son of Zeus". MNZYVZ (pronounced Nevis) is the RTTUR figure, one of THE TWENTY FIVE, not Zeus.

The Ancient Greek Roman and other many God Systems have the MAN MADE understanding in them which is INCORRECT.

RTTUR, THNGME, to whom the other TWENTY FOUR RTTUR defer, is extraordinarily difficult about his name

Much of the difficulty is that local, teaching and dogma offends the Galactic position.

It is all one to say and record that Saul "saw the light" on the road to Damascus, but in real Galactic terms, the veracity of that is suspect. Paul as Saul had been involved with the murder of Stephen, and was thus not someone who would be likely to find favour with these vast Galactic civilisations, for whom murder, especially judicial (and that does include suicides in custody) is regarded as The Abomination of Desolation. In their terms SYNA (Sin) and thus puts into serious question much of New Testament Text after Acts, and before the Technical Specification which is Revelation.

One important aspect with the Galactic Science is that there must be freedom of expression to criticise, even that which appears sacred, and where that is curtailed there is no prospect of Galactic Enlightenment.

Matthew does not cause any difficulty, because it is M A T T HEW, which I have explained is the Version provided by the Galaxy Systems, MNZYVZ AAGHAADDWR THUURRNN THNGME THRENZ.

I am also aware that the spelling hides considerable technical detail. Thus U V +W , when used are for secured Galaxy Systems (00) as U, and W as (0000). In the case of THUURRNN I could write as TH00000000RRNN, but that is cumbersome, so TH08RRNN could be used instead. That is also cumbersome, so TH0RNAN is an easy usage which hides the technical detail; unless it known what to look for.

In many respects it is all to do with pronunciation, and that is easily misunderstood by the locals, even if time is taken to explain.

Thus ANNUNAKI can be written as ANN00NAKI, and once that is done the (00) discloses the individual as Galactic. Ky is masculine, so I have the Director of an 00 system, as with AH00RA MAZDA where the MAZDA is for "The One Galactic Understanding".

The names may change but the underlying DRAGONSPEAK Super-communicate still remains.

There is some suggestion that Moses lifted some of his material from other location's writings, before he set about writing what are the first Books of the Old Testament. TH 0 RA could be said to be a reference to a White Tyger Headed THNGME, Head of the TH0RNAN GROUP GALAXIEZ, and the number ONE of the TWENTY FIVE RTTUR!

Thus the criticism by Jesus, that what the locals were working to, was man-made rules and is exactly what he would say, where he had the Galactic Understanding, given to him both from his time in Egypt but also from THE KHRYZTAAL GALAXIEZ who had sent him to run their tests. The difficulty there as disclosed by the Matthew Text is that he was a KURALAKYZ individual, with a Galactic connection; vastly different from any Pharisee or Sadducee.

In writing this text, as I indicated in Book Six, The Lost Atlantiz, Moses was annoyed that he had only very limited access to the Galactic Science. No surprise there as he had killed an Egyptian, and in his texts had put in place judicial murder and mutilation. It was thus a considerable gift that he was taken up a mountain to see THE FIRE THAT BURNS AND BURNS AND DOES NOT CONSUME; part of the large Spiral Disc which is this Galaxy ATLANTIZ_DRAAZ, monikered as THE GREAT BLUU WHAAL, with the necessary PYKAAZ_PYNWHAAL identifier of the whale. The sign of the fish is only partial, account must be taken of the THE GREAT SOUTHERN PYNWHAAL Galaxy and NORTHERN PYNWHAAL Galaxy and the technical species design "Milky Way", which is male and female. The Design aspect is Critical; thus the first words of Genesis 1:1. Yes there was an attempt to put the correct design in place, but as should be obvious from the Starry Sky, that is a Galaxy matter.

My attention was drawn to HABAKKUK, and its short style discloses technical Galactic Science. [To communicate, brief words are best.] I have to remember that B & V are similar in pronunciation, and as V is often used instead of U it is an identifier for a hidden and hugely secured Galaxy. Thus change to HA VA KK UK and with UK changed to 00K I have again a very large group masking off as Hidden. With AT 0N and 00K I have hidden Galaxy markers, but well known to MR DRAGON ZYR.

Hidden Galaxies are thus Dark Energy and matter, because as they spin there is considerable energy involved, and energy is required to MAKE MATTER. That is correct. I am talking about making matter out of energy. That is what building and manufacturing does.

The blue print is design; the energy is expended in building and manufacture.

In these days when ZER0 is worshipped at the expense of everything else and dogma is worshiped as Truth, where that Dogma is fundamentally a LIE, there is very little scope for the Real Galactic Science and the REAL Galaxy RTTUR can withdraw and lock down.

The "Burning Bush" and the so called "World Tree" deal with the same Galaxy understanding and as I have indicated previously YYGDRAAZAL is a concealed term for ATLANTIZ_DRAAZ Galaxy.

Thus in many respects THE TRUTH has always been there (which fundamentally is the definition of Truth which does not change) hidden over the millennia by spurious Dogma.

If I am looking for Thor's Hammer, I need look no further than the destruction of Hiroshima and Nagasaki; and to the Galactic input there, for the Nuclear weapons are DRAGON'S

TEETH, and the local Dragon, hidden in plain sight(now you see him now you don't) is the Local LANAZT STAR.

So with my texts I have updated and explained some of the Ancient Dogmatic Mythology.

The RTTUR have pointed out, that the PROOF is with the term YYGDRAAZAL, as a very concealed reference for ATLANTIZ_DRAAZ; which is this Galaxy.

The nine realms of Nordic mythology are simply an attempt at describing NINE GALAXIEZ, to which it is possible for some to travel. My spelling of FVHAALYA, which is the principal city in FVREKKRAN, M51 Galaxy, is thus the location of one of the REALMS; a disclosure matter, to be agreed with RTTUR FVRRAKOTHERA.

I wish to travel from Glasgow to Invercargill. My journey has to be largely by plane(as with the Boat of Millions of Years), so I travel from one Realm, Scotland, to another Realm, which is England, next stop is Perth Western Australia, another Realm, although it might be said to be the same Realm, then to New Zealand; which is another realm! Depending on the view taken of "Realm" it is all in the same realm; as within this Galaxy.

In Galactic Science terms I can replace the names with Galaxy Names, as my Star Charts demonstrate. Thus ATLANTIZ_DRAAZ to PP HAMA GIGHA, to FVREKKRAN to LLYKOTT_LLYDRAN to FVRGHAAL to MNZYVZ to THUURRNN_ THNGME; which could be said to be the "Seventh Heaven". If there are nine Realms, I need to take in another Two Galaxies, which would be PYNWHAAL and AAGHAADDWR and I thus have NINE GALAXIEZ.

The reality in question when talking about ATLANTIZ_DRAAZ are the FOOTSTOOLZ in this Galaxy accessed by means of Wormhole Technology(thus the reference to the double headed serpent.) where they are weighted at 1/- plus. That is a very advanced mechanism, and with this location at 3d Ag 500, not at a stage to be able to utilise that travel mechanism.

Rocket science, as I have said previously, fine for satellites, but no use for serious Galactic Travel to very advanced civilisations.

V

The World Tree

This is a description from Norse Mythology. The problem with the Mythology is that it is rooted in fact and reality, but often very difficult to appreciate unless the correct understanding is known.

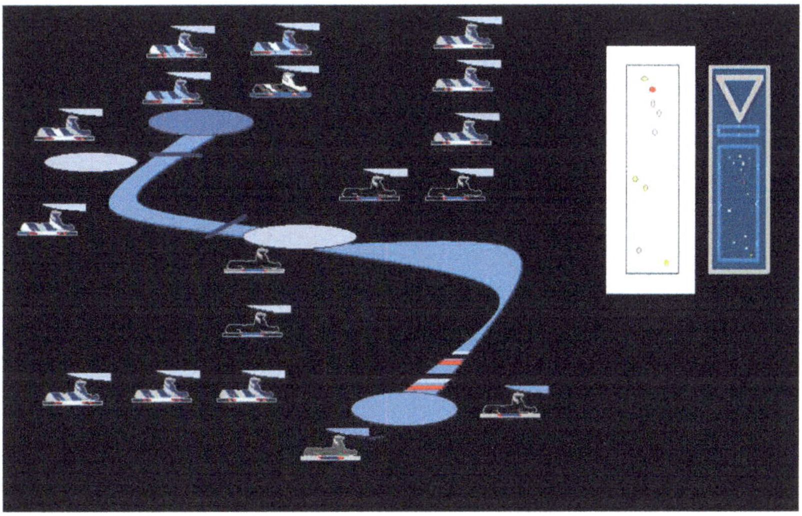

I refer to this as a Galactic Star Chart because that is what it is. With its black background it is a hugely secured chart. The bottom blue disc with its two Sphinx(Resting Tyger) light grey for ATLANTIZ, Black for DRAAZ represent YYGDRAAZAL, and it is a representation of the so called "World Tree" because rather loosely it resembles a tree. The start off point, the "roots", are at the base which with the barring shown, means it is thus this Galaxy, with its hugely secured name.

From this Galaxy there are "Gates" which are shown by the red blue and black lines across the route, and they signify a barring as there is a lack of correct Galactic Understanding. Excluding the black there are nine lines, so that signifies keys which have to be obtained.

The next destination represented by the circular disc is the nearest Spiral Galaxy in System, which is THRENZ. It is guarded by Ten Resting Tyger, of which 4 are very advanced £1.+ 925 Ag minimum, and 6 at 1/- level 925 Ag. That means that TEN stars in THRENZ have to give clearance for entry. That might be said to be Ten Commandments which have to be correctly understood. Where the travel is to other Galaxiez, The Ten Commandments mean that TEN GUARDIAN FOOTSTOOLZ and their STARS have to provide the keys to allow inter Galactic Travel.

The light grey Resting Tyger is ATLANTIZ, which does not have the technology for super light travel.

On the route the next Disc is not shown in the route, but is blacked out by two lines, and it has two Resting White Tyger. It thus has Superlight travel, with matter transfer but is not a stopover.

The top disc has four Resting Tyger, all of which are at 5/- Ag 925 or above and is the destination Galaxy for travel. The Tyger give an indication of the Galaxy involved and that it is shape shifting in approved style, ATLANTYON and TYGER. It is FVREKKRAN, and the destination city will be FVHAALYA.

It will be noted that although the spelling has changed the pronunciation is similar.

In this Chart there are only three Realms. YYGDRAAZAL, THRENZ (the one bye passed, which is PP HAMA GIGHA) and ending up at FVREKKRAN.

There is a compulsory stop over after three Galaxies and another star chart, of similar design would be required for the next part of the route. There are twelve pilots on a Boat of Millions of years, because some GALAXIEZ, for security, insist on their own pilots. The next chart could be from FVREKKRAN to AAGHAADDWR to TY MA HY HAAT, and then a further from

TY MA HY HAAT then MNZYVZ to THNGME. Only certain routes can access such as THNGME, with certain stopovers.

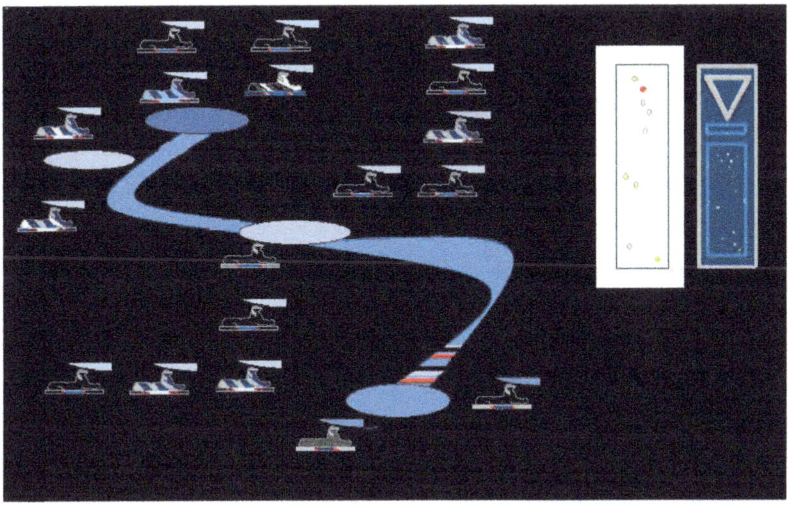

The resting Black Tyger represents HWRKHWART UMZAARAN (Hugely Secured) FOOTSTOOLZ.

The departure point in this Chart is from the Black Resting Tyger,(lowest right) and in a Galaxy which has others at less than the 1/- weight.

A very similar chart could be done for just ATLANTIZ DRAAZ, and in that case with a 10/6d Galaxy weighting, it is DRAAZ at 2/6d Ag, one other with superlight at 1/-, and twenty eight of the pale grey, at 3d(which should be minimum 3d Ag , all 925.)

The only difference between an Intergalactic chart and a local Galaxy chart, is that the inter Galactic has a Dark blue or Black background, and the local Galaxy is one which has a light blue background.

The black Resting Tyger in YYGDRAAZAL indicates that at least one Star and Footstool, has Galactic links.

Once this understanding of the "World Tree" is lost, the Galactic Links are down.

The "Burning Bush" is a similar Galaxy description, usually reserved for this local Galaxy. "The Golden Fleece" is another for this Galaxy, but where the YYGDRAAZAL term is used the "Moniker" for this Galaxy, to highlight galactic connection with PYKAAZ and PYNWHAAL is "The Great Bluu Whaal"; with the different spelling highlighting the Galactic Understanding.

The Galactic Understanding is Galactic DESIGN, with RTTUR individuals. The TWENTY FIVE ELDERZ if that terminology is used.

I gave in the technical specification of The Lost Atlantiz the various RTTUR names, with male and female notations for heads of Galaxy. Given that they can shape shift to the four designs ATLANTYON, WOLF, TYGER and RHINO, it is a matter for each RTTUR to determine in which style they appear in from time to time, if they choose to place their feet on a FOOTSTOOL.

The local population can do their own mythology, which can be very wide of the mark, but has to contain the essentials of the compatible Galaxy; A HEAD RTTUR and "Milky Way Design" male and female.

The difficult aspect with RTTUR is that the names can change, and that again depends on the level of understanding and particularly of enlightenment.

If a local population is NOT ENLIGHTENED, this System Group will only have very limited contact, the so called "thief in the night" exercise.

From a distance they can easily see if their PROTOKOLZ might be met, but it is only UP CLOSE and personal that the position can be checked.

To assess compatibility, the Local Star has to give assurances that the PROTOKOLZ can be met; and will involve a minimum of 70 years, and usually four Generations of 70 years.

The RTTUR assessment as of 2024 is that with reluctance 3d Ag 500 is given. The criteria to be met are (First) safe civilian air travel, and on a time distance London to Auckland of 10 hours.

That is all hugely technical; and you cannot have aircraft blown out of the sky by terrorists or missiles, or deliberately crashed into mountains or into the oceans by pilots; and that would include by accident and (Second) local societies which are open and free (freedom of Speech being a critical assessment) and the assessment point on that is the Pontius Pilate matter; on which this current population continue to fail.

They only reason these advanced civilisations take any interest in this location is because it is one of their FOOTSTOOLZ in this ATLANTIZ_DRAAZ Galaxy. They provided the NYKYDYNYZZ design material MALE and FEMALE, with reproduction in "the old fashioned way." That aspect of design is material, and "pea soup and evolution" is wrong Galactic Science. Such thinking is a BAR to travel in the Galaxy, because the species are NOT COMPATIBLE.

Various old Mythologies thus contain elements of the Galactic Science, but not in a sufficiently enlightened format, the mistake commonly made locally is to try to force the Galactic reality into a local dogma. Square peg to round hole!

The New Testament text was an attempt at enlightenment, which was refused by the locals who insisted on their local dogma. Again in 2024 the insistence is on local dogma, as typified by the drive to no carbon and State Censorship, with vindictive State Security. That is not compatible with Galactic Science, and thus the 5,101 year LOCK DOWN is correctly applied. It is a LOCK DOWN matter.

Although only dealing with a small local area, the test is both GLOBAL and GALACTIC.

Only those civilisations which can do matter transfer can change designs, and the R18 Protokol, is insisted but not forced. Failure to observe that is a LOCK DOWN matter, and indeed would extend to loss of even the 3d Ag status. On that the RTTUR could instruct snapping of the DRAGONSTICK on which has been placed the up-rated Design FLYFLOTT for ATLANTIZ_DRAAZ.

RESERVE AND REDAKT on that!

The designation AT 0N is likely to be removed, and that as with most LOCK DOWN matters is done locally. Locally the letters have had the order changed, but they do reflect the ENORMOUS DRAGON, which is the ATLANTIZ_DRAAZ Galaxy.

vi

The words RESERVE AND REDAKT mean that there is no free speech in the Galactic Science terms and thus also incorrect science. That is typified by Dogma preferred to Truth. Further LOCK DOWN MATTER.

vii

BASICS of GALACTIC UNDERSTANDING

GALAXY NAMES AND RTTUR NAMES

These are DRAGONSPEAK terms and they vary considerably. **There is a loose connection to local names, Thus "Whirlpool, [M51]" and FVREKKRAN. It is FVREKKRAN and their RTTUR** who has to agree to the release of their correct GALAXY and RTTUR name. Different spellings identify different levels of **understanding. This text is at "GREAT BLUU WHAAL" level,** which is for a Real Galaxy with Real Galactic Connections, 2/6d plus, even where it is currently LOCKED DOWN and the FOOTSTOOL weighting is only 3d Ag .500.

Different Galaxies have their abbreviations (somewhat similar to airport destinations locally).

Thus the local population must accept that design is Critical, and ensure that Truth and Science correctly combine.

Scientific Truth is required for basic Galactic Science. Thus it is necessary to correctly understand chemistry. It is not required that everyone understands to the correct in depth level, but that such as the periodic table is available to be correctly understood.

Biology comes under similar requirements, with the basic requirement being male and female, including the animals, plants and water living creatures. Again there are a few exceptions, but the male and female in strict definition is important.

The Galactic science aspect is with design of Species.

It is fine for those dealing with Science fiction to vary the styles of alien heads, and indeed for the bodies, but as this is a KHRYZTAAL GALAXIEZ FOOTSTOOL, THUURRNN only permit certain designs.

The ancient Egyptians were good sculptors, and thus they were able to create the NYTYR designs. There is a trick here that only some of the designs work, others do not and that is done to make the point that it is only certain restricted designs which are permitted. Colour is very important. Red is a SEKHMET Colour and she is mostly DRASSYAN. Red Tyger can be taken by any of the other Galaxies, where they have Tyger Headed species. However the names will usually provide the Galaxy reference.

The dog headed figure of Anubis is a correct design, but in system Galaxies designs, there is a triple style grouping, thus RED, WHITE and BLUE. It is not RGB.

The black is in fact midnight blue, White is effectively "pink skin", and Red may vary to light orange (yellow.)

THRENZ has black, white and Grey TYGER of the bi pedal style.

LLYKOTT_LLYDRAN has black, white and red including in the rhino style, bi pedal and in Atlantyon

Dark Blue, Pink skin and light orange.

These are design matters of considerable importance. They do not change.

When I am presented with UFO photos of dead Aliens which do not conform to these specifications, the only term which applies is FAKE.

In connection with UFO's there is a huge level of fakery which simply demonstrates that the locals here do not have the correct understanding. Even where the Science fiction is examined, the knowledge of superlight travel is incorrect. It works from KNOWN DEPARTURE to KNOWN DESTINATION; in other words from an advanced civilisation 2/6d Ag + to 2/6d Ag plus.

Occasionally those above 5/- + Ag 925 may visit such as 3d to check out the local position, but such can do so using the local internet where the society has such, without the need to travel.

The travel is usually by Boats of Millions of Years, which are specifically designed to hide the advanced technology of superlight travel and NOT TO FRIGHTEN the natives.

The distances involved are security barriers, thus THRENZ (M31) at 2.2 million light years. Others are further away.

Generally it will be an RTTUR or a KURALAKYZ who will use the boat to travel. The Boat is the property of a specific RTTUR and the KURALAKYZ working to the Direction of the RTTUR.

Thus Zarathustra, the KURALAKYZ GYDRAN to the RTTUR for AHURA_MAZDA Galaxy. Of course where there are several Galaxies in a particular group, a lesser Galaxy may hide a more powerful. Thus NYMRRYDD, is in the MNZYVZ grouping, and it is important to understand that the most powerful are often carefully concealed behind others, and that includes with their names.

HWRKHAART UMZAARAN THNGME falls into that category, where their RTTUR would be described as I AM THAT I AM; carefully avoiding reference to any of his specific names, especially LORD THNGME.

To put that down in print here, I need the approval of RTTUR THNGME, even when he is in Distant THNGME GALAXY in TYGER FORM, which he likes, in the City of KHUUMN.

The MN start of a description or at the end signified THUURRNN connection, to that group of 1000+ Galaxies, most of which are securely hidden from prying eyes.

So "EarTH" is one of THNGME footstools, and the local LANAZT Star is to largely THUURRNN design, with THUURRNN plasma.

The basic Galactic understanding is that the Star and Footstool are linked and that the mechanism is a gigantic distillation plant, to provide for the species.

The gravity is determined by mass and spin speed, and the volcanoes HYHASTERN, inlet valves, and KULYN are the exhaust valves used to regulate the spin speed.

The balance weights are the ice sheets, north and south, which are melted and frozen to reposition as the local STAR and the SYSTEM GALAXIEZ require.

Like a large train turntable, the wormhole links are only available when the globe is correctly positioned to line up, and that lining up may be referred to as a "Giant's Causeway."

REVIEWING ATLANTIZ SOURCES

I am satisfied that my understanding of ATLANTIZ in the Galactic science sense is correct; particularly with the concealed term YYGDRAAZAL. I am aware that the Nordic Spelling for the World Tree differs, and I am obliged to check the Nordic information. As usual I must have it in English Language format, but that I adjust. The mischievous LL0KY with his shape shifting abilities draws my attention to LLYK0TT_LLYDRAN, and RTTUR LL0M0TT0R. Again changing the "O" to "0" to represent the Galactic Discs, I have a triple Galaxy Combination. Is then LL0KY the principal individual neatly hidden behind 0DYN and TH0R? TH0R is clearly THUURRNN Galaxy, and the Triple

Galaxies are in those myths as Giants; and of course the Galactic Discs are very large.

That brings me to PLAT0! Perhaps a change to PP LAT 0 to mean the Compatible PYKAAZ_PYNWHAAL, Milky Way Species type of this vast Galactic Group, I call THE KHRYZTAAL GALAXIEZ.

Plato does not finish his story, for the simple reason that the GALAXIEZ will not permit it at that time. Greece and much of the Middle East is in LOCK DOWN because of serious Malpractice. A matter I can check out in H0SEA and the other short books of the Old Testament where "0" and "Z" appear as markers for correct Galactic Understanding.

For my part I am very conscious that while six or seven books make a lengthy book, each one is relatively short and may be referred to as a "little book." I delay to authorise other than a limited issue of Book Six, The Lost Atlantiz, because the locals have lost ATLANTIZ. This one is likewise restricted because book Six is referred to.

viii

ACTIVATORZ

I referred to these in a chapter in the Old Text here

In determining assessments, in terms of Daniel 12:12 certain events have to take place before these RTTUR will instruct a SLAP DOWN.

Each RTTUR has their own dislikes and to authorise a SLAP DOWN all TWENTY FIVE have to mark NOT KONTENT.

The most obvious for NOT KONTENT are vindictive legal rules. In DRAGONSPEAK this may be masked off.

As of this date because of local Rules I am unable to clarify and it is a RESERVE AND REDAKT matter but as of 1 April 2024 AD 107 all TWENTY FIVE RTTUR have marked NOT

KONTENT and LOCK DOWN IS IMMEDIATELY INSISTED as of that date.

The name, Baroness 0RCZY makes the point, as did Herman Melville with **"the White Whale."**, as if to clarify a Galactic understanding.

It is not the hunting of the Snark, although I suppose a Galaxy **might be referred to as a Monster, but rather the "mad hatter"** with his 10/6d price ticket, when 10/6d turn out to be the 30 pieces of silver which make up the minimum Galaxy weight for a SYSTEM GALAXY FLYFLOTT.

I may be said to be stretching the connections, but that is exactly, or should I say, PREZAKTLY, what the DRAGONSPEAK requires and is INSISTED.

Like the Norse myths, the DRAGONSPEAK is there, it is simply very difficult to see. LL0KY is at work! A female horse indeed, and a horse minus cajones, is still a horse of male design. That makes Loki even more difficult, but I am well aware that LL0KY is insistent on rampant males, but with females treated with respect. What does LLOMOTTOR make of show trials on sexual matters minus a jury? I spy a NOT KONTENT. Likewise THUURRNN will also deliver a NOT KONTENT on that.

The locals here are not interested. They do their own thing. Dogma insisted and preferred to Galactic Science.

I stand in Candalaria in Tenerife, which is a rather large volcano, where the rim of the caldera is all the islands which make up the Canaries. Not a small volcano, but enormous. The front cover photograph is of a view to La Gomera from Tenerife.

My ATLANTIZ is the entire Globe, but on that Galactic Interpretation it has to have advanced enlightened individuals in Atlantyon style as with ANDRUAITH and PYTK , who withdrew back to their own Galaxies leaving very little trace of their presence. This location is thus THE LOST ATLANTIZ.

The Norse Myths and Plato make reference to a serpent, which is a sure indication that it is the super light travel mechanism which is involved and not correctly understood.

If the local population, or large parts of it, do not understand correctly, are not taught balance and fairness in the legal system, with no harsh and oppressive penalties, if they do not observe such as the Hippocratic Oath and indulge in wrong medicine, what have the local population to offer to these Galaxies?

I sun myself on Tenerife and am happy there. Now I am unwelcome there. All technical as LLOMOTTOR understands. I am made unwelcome and threatened here in Scotland as those in charge seek to abolish balance and fairness in the Legal System and bring in harsh and oppressive rules. I have two State murders of under 18's in Scotland, caused by harsh and oppressive penalties. Am I not entitled to chuck the King James Version at them at utter disgust at their corruption? The European Convention on Human rights is *de facto* shredded by the locals. That is a minor matter. Upsetting the GALACTIC RTTUR of THE GREAT COUNCIL is not prudent.

Two white Tyger restrain me. I am not to push the button. Let 2000+ years go by and do a review. It is LLOMOTTOR and LLYMOGREN who restrain me. I am not to play THOR and do the slap down.

RA PA NUI. Do it not! An Atlantyon phrase as politely put to an RTTUR.

So I have to check the Maori myths, and there I find the shark as a reference for the "Milky Way." But "Milky Way" is a species design, male and female, and New Zealand is in the territory of The Great Southern Pynwhaal Galaxy; MR WILD ORKA ZYR to use that moniker; RTTUR FVRGHAAL. Another giant!

So in essence even with Plato and many other philosophies, the Galactic Science is there but it is simply not correctly grasped and put together.

It is not for my Star Charts to fit the nine Realms of Norse Mythology, but for me to explain the Galactic position and style.

That I consider I have done. Such Galactic Star charts are usually on a Black background; although for some it must include at least a touch of blue so that it is not pure black.

A similar chart, with a blue or light blue background is only for this Galaxy.

The design is basically the same. Four Galaxy Discs are shown, three for approval to travel superlight within this Galaxy. There is an insisted break on reaching the destination FOOTSTOOL, and a new chart would be added, particularly if it was onward travel to another Galaxy. An open Star chart will not show the barring. The barring is at the start, where a location, as with ATLANTIZ does not have the correct level of understanding.

These are TRIPLE GALAXIES charts, and the four Discs represent the SYSTEM GALAXIEZ.

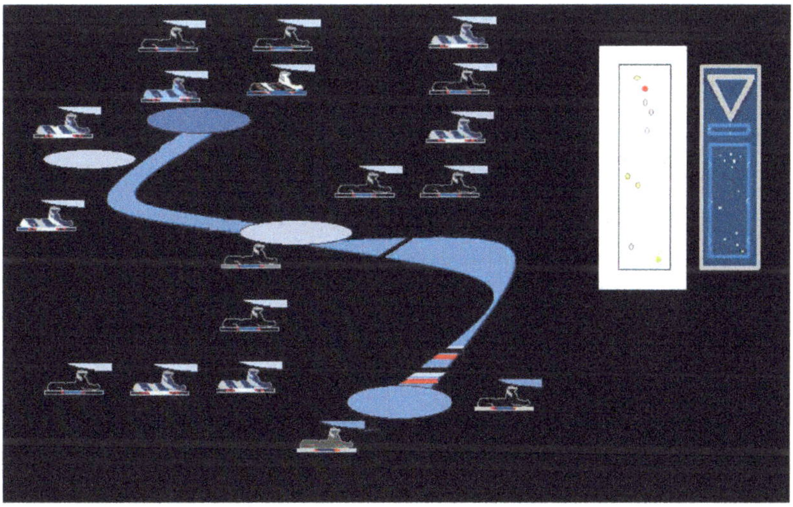

Because this Galaxy as ATLANTIZ_DRAAZ is connected with TRIPLE GALAXIEZ any superlight travel has to be agreed by the two others in the route, and depending on the destination, may also require approval from the one not in the route. The black background here is an ADVANCED Chart, but makes the same point that ATLANTIZ is closed to the super light travel mechanism because of the barring before the nearest Galaxy, THRENZ, M 31.

Nearest is M31 THRENZ, which is the middle blue disk with TEN Resting Tyger Guardians.

The top portion includes the Third Group, represented by the group of four resting Tyger.

Assuming groups two and three agree, travel is then within the bottom group which is this Galaxy as ATLANTIZ DRAAZ.

The dark grey resting Tyger, bottom left is this FOOTSTOOL (with the dark grey representing 1/- equivalent Ag .925 +, and the Black representing DRAAZ, within this Galaxy. All the Black resting Tyger are at 2/6d or above. That means that in addition to superlight, they have correct matter transfer and transiting. They will all have to agree the travel.

The White and Grey are the more restricted FOOTSTOOLS, which can only be accessed, from the Black.

The charts look primitive, but as should be apparent to anyone who uses such as a modern tablet, such tablet can be used to select the shapes as destinations.

Punch in the detail, starting at bottom left dark grey, as ATLANTIZ, then next punch in destination, which is the bottom Right which is DRAAZ. Thereafter there needs to be punched in another two black, and the names must be correctly given.

The chart shows FOUR other advanced FOOTSTOOLZ in this TOP GALAXY, which makes two at 2/6d+, 2 at 5/-+, so taking the galaxy above 10/6d to make it a SYSTEM GALAXY.

In fact it is for DRAAZ to give the clearance, and it will do that by checking with M31 THRENZ and one other in the top group (could select , PYNWHAAL, AAGHAADDWR, MNZYVZ or LLYKOTT,) to give clearance to approve the super light travel from ATLANTIZ to DRAAZ. IT IS NECESSARY TO KNOW THE NAME OF THE PRINCIPAL CITY. Thus if MNZYVZ is to be asked for clearance TRYSON ON AN AP AN APIZ has to be given. That can be shortened to APIZ, but since all of these are

PASSCODES, further PASSCODES have to be correctly added. Incorrect PASSCODES no travel; and LOCK OUT. Could be for hundreds or thousands of years!

In the first instance it is only the short journey of some 15,000 light years which would be approved, to enable some passengers to be taken from here at ATLANTIZ to DRAAZ, and with recording equipment so that they might take video of one of the large TYGER cities; most likely Taydaranakushton(which would make New York and Manhattan look tiny).

DRAAZ has to give approval, but it has to check with other Galaxies as it has links to them which are open and active.

For the most part the BOATS of MILLIONZ of YEARS are used by the RTTUR and their KURALAKYZ to check out a footstool, as to whether or not it is safe to "place the feet." An RTTUR will generally only visit at the very advanced FOOTSTOOLZ, leaving it to a KYURALAKYZ GYDRAN to do initial assessments of those FOOTSTOOLZ, which have a weighting of less than 2/6d.

Where the Star chart is viewed as the base of the "World Tree" that is light grey AND black Resting Tyger, where the Black Resting Tyger represents DRAAZ, and the combination with ATLANTIZ now agrees to the superlight travel.

The middle disc represents THRENZ, and one of the black Resting Tyger Guardians is HYATT GHARVIK BY AT RUTH which have to give clearance to enter THRENZ, or to obtain clearance from them, even where the Galaxy route is from ATLANTIZ to DRAAZ, within this Galaxy.

It is very much NOT a matter of trekking into the unknown, but to know PREZAKTLY where you are going, complete with the necessary clearances (Visas) and after full medical health checks; as scanned and rectified so that the PROTOKOL, "Nothing which is harmful or hurtful will be admitted," is met.

The system is heavily Guarded with many more predominately Tyger headed populations rather than Atlantyon.

The black resting Tyger in this Galaxy are likely to include RHINO (including red), WOLF(which would include such as Anubyz [AN 00 BY Z] style) on FOOTSTOOLZ in this ATLANTIZ_DRAAZ Galaxy.

The double headed superlight travel mechanism means that to travel to DRAAZ, DRAAZ has to open the Galactic Gates here, and the Gates in DRAAZ have to be opened from here. THAT must be done safely and without injury; a very material matter, which is the obvious mistake by 0DYN in the Norse Mythology, and to a certain extent, the technical mistake in the New Testament Text.

How many times does it have to be said, NO HUMAN BLOOD SACRIFICE? NO TO SACRIFICE OF ANIMALS for religious observance either! Animals can be used for food, but due regard must be given to animal welfare.

Sin, SYNA in Galactic terms is the worship of murder and mutilation; THAT is the LOCK DOWN and THROW out matter!!! The local star MR DRAGON ZYR, is well aware that when that is done by States, Destruction by FIRE is authorised; thus Persepolis, Hiroshima and Nagasaki.

The specification is in Daniel 12:12. 1335 days has to expire after the attack on PEARL HARBOR (any spelling provided it is understood as a System Footstool.) It is a matter for MR DRAGON ZYR; not something HUMMINGBIRD will usually permit.

Nothing which is hurtful or Harmful will be admitted. Final Answer! It is INSISTED PLUZ Y PLUZ.

ACTIVATING A GIANT'S CAUSEWAY

It is a NO GO for those who insist on blood sacrifice; for such it is fatal and they locked and barred from entry.

LLOMOTTOR can take some if he wishes, but such are barred from the THUURRNN Systems. That was the position at the beginning. MNZYVZ can check on remodelling, but it is doubtful.

It is the RTTUR and their KURALAKYZ GYDRAN who visit the footstools to apply the grading. There is no way this location can move from 3d Ag 500. The level of injustice stinks to Heaven. The discourtesy shown to the KHRYZTAAL GALAXIEZ officials is simply appalling. They cannot stick corrupt Governments, and **having provided many Dragon's TEETH it may** be playtime for MR DRAGON ZYR.

The local LANAZT star asked for the review, and it is done. 3d Ag 500 is an improvement, but seriously **via veritas vita** it has to be. Nothing else is acceptable. THE TWENTY FIVE will NOT agree to a different standard.

The little book is easily done. Plato's Atlantis weighs in at 196 Grams, The Norse Mythology at 292 grams, and my text of The Lost Atlantiz at 335 Grams.

10/6 Ag 925 weighs in at 60 grams, but in the combination 1 x 5/-, 2 x 2/6, and 2 x 3d all Ag 925, gives me a Galaxy, with three extraordinarily advanced FOOTSTOOLZ, and 2 which are only at a similar development to this one at AD 2024 107.

The Norse Mythology was provided by LLYKOTT_ LLYDRAN, M64, which is in the FVREKKRAN grouping, my LOST ATLANTIZ text, represents THUURRNN, and in particular THAT THNGME Galaxy, NGC 1300, and that leaves PLAT0 with this Galaxy, with assistance for PP LLYAAZ_ATTHENYA, FVHATTHARAN group, and of course assistance from THNGME at present.

I had wanted to take in a train ride from the City Centre palace of the Lady Kyen Ty Amen Ty to her summer palace in the mountains, a location a bit like New Zealand, but the Lady Kyen will not agree to the release of information, save to say it is in a continent similar to and at a location like Australia and New Zealand combined without the Tasman Sea in between.

Like PLAT0 I may have the wrong understanding, but I have tried to clarify. I am aware that the KHRYZTAAL GALAXIEZ GREAT COUNCIL restricts use of and clarification, save to make it clear that their KURALAKYZ YYG ZYVZ, TYEGA TZUTRAN ADATRAN, TYEGA ZYON is a GALACTIC individual who should not have been threatened, far less judicially murdered. MNZYVZ is insistent on the apology, and will not be placated until he has that; notwithstanding the resurrection of his KURALAKYZ.

MNZYVZ is still grumpy and does not do forgiveness.

THNGME has just had his selective amnesia fail.

LLOMOTTOR (LL0KY) is just waiting to release his Red Rhino to do the tidy up; the plagues.

<<< SAID LLOMOTTOR to LLYMOGREN >>>

"The position is accepted and has been conveyed to MR DRAGON ZYR and to HUMMINGBIRD.

It is FIVE NOT KONTENTS; ACTIVATION on 1 APRIL 2024 AD 107 INSISTED"

ix

TECHNICAL NOTE ON WEIGHTING Ag

FLYFLOTT WEIGHTING

The FLYFLOTT is simply the symbol of Correct Galactic Understanding. It may be hand carried, with the cords to hold it in place. That is what the NYTYR "ANKH" symbol represents. KH "KYDRAN or KURALAKYZ HA (Enlightened). The AN is for ANANAN (meaning living- A real individual) The FLYFLOTT and Headdress is thus a symbol of Galactic Enlightenment of compatible individuals.

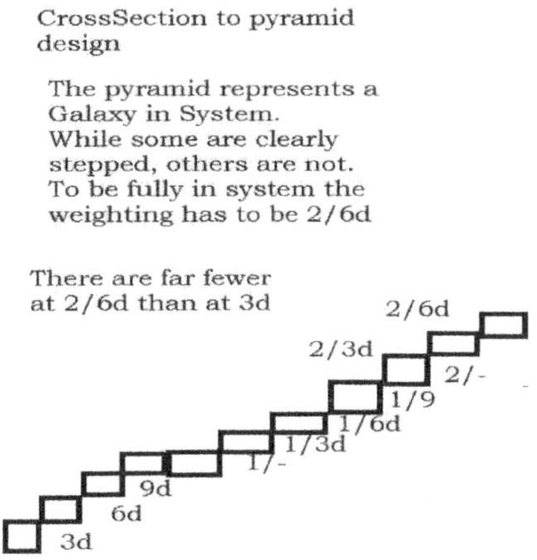

As drawn here in profile it might be said to represent a ladder, as steps to "Heaven", where "Heaven" is at the 2/6d + Ag 925 FOOTSTOOLZ.

Ag at purity .925 and above represents a FOOTSTOOL. Where Ag .500 is used, the value of the Weighting is halved.

3d Ag 500 civilisation has basic Civilian Air Travel and medicine which is Hippocratic Oath compliant. Level too low for other than restricted contact.

3d Ag 925+ civilisation has basic Civilian Air Travel and medicine which is Hippocratic Oath compliant.

6d Ag 925+ has an acceptable, balanced and fair legal system which does not indulge in harsh and oppressive penalties. No show trials.

9d Ag 925+ As with 6d Ag 925+, but this has be the position for at least 500 years

1/- Ag 925+ A society which has superlight travel, but not in an acceptable style.(That would include unsafe local Legal and medical systems).

2/6d Ag 925+ A society which has superlight travel, with matter transfer, and transiting in an acceptable style.

5/- Plus Ag 500. This is equivalent to 2/6d Ag 925+

5/- plus Ag 925+ As with 2/6d Ag 925+, but has maintained such over at least 1000 years.

On the Star Chart there are two rectangular boxes on the upper right; one shows the route of travel, the other shows the levels within the Galaxy. While the Galaxies may be viewed like plates, some like TY MA HY HAAT resemble a "Sombrero" or "Fedora". The Galaxies have, in the case of Discs, a Galactic Plane with levels above and below the plane. Such levels are important in assessing whether the direction of travel in the Galaxy is level, uphill or downhill.

x

This photo is to show a KHRYZTAAL GALAXIEZ style plate (bowl) explanation for such as TY MA HY HAAT [Sombrero]; a very advanced GALAXY. The weighting in the KHRYZTAAL glass is £2+2/6d, with coins all Shielded in plastic cases; meaning they are hugely secured FOOTSTOOLZ. The 3d bits in case, total

2/6d so that all FOOTSTOOLZ on bowl are at 2/6d or above. The design of the bowl and mat represent PYKAAZ PYNWHAAL design. It is the Ag in ATOMS which are relevant. The UK coins are used for convenience.

METHINKZ THAT WRAPZ IT UP!

Made in the USA
Monee, IL
02 November 2026

67385872R00085